FORSCHUNGSBERICHTE AUS DEM LEHRSTUHL FÜR REGELUNGSSYSTEME

TECHNISCHE UNIVERSITÄT KAISERSLAUTERN

Band 4

T0135721

Forschungsberichte aus dem Lehrstuhl für Regelungssysteme

Technische Universität Kaiserslautern

Band 4

Herausgeber:

Prof. Dr. Steven Liu

Jens Kroneis

Model-based trajectory tracking control of a planar parallel robot with redundancies

Logos Verlag Berlin

λογος

**Forschungsberichte aus dem Lehrstuhl für Regelungssysteme
Technische Universität Kaiserslautern**

Herausgegeben von
Univ.-Prof. Dr.-Ing. Steven Liu
Lehrstuhl für Regelungssysteme
Technische Universität Kaiserslautern
Erwin-Schrödinger-Str. 12/332
D-67663 Kaiserslautern
e-mail: sliu@eit.uni-kl.de

Bibliografische Information der Deutschen Nationalbibliothek

Die Deutsche Nationalbibliothek verzeichnet diese Publikation in der
Deutschen Nationalbibliografie; detaillierte bibliografische Daten sind
im Internet über http://dnb.d-nb.de abrufbar.

© Copyright Logos Verlag Berlin GmbH 2011
Alle Rechte vorbehalten.

ISBN 978-3-8325-2919-2
ISSN 2190-7897

Logos Verlag Berlin GmbH
Comeniushof, Gubener Str. 47,
10243 Berlin
Tel.:   +49 (0)30 / 42 85 10 90
Fax: +49 (0)30 / 42 85 10 92
http://www.logos-verlag.de

Model-based trajectory tracking control of a planar parallel robot with redundancies

Modellbasierte Trajektorienfolgeregelung für einen planaren
Parallelroboter mit Redundanzen

Vom Fachbereich Elektrotechnik und Informationstechnik
der Technischen Universität Kaiserslautern
zur Verleihung des akademischen Grades

Doktor der Ingenieurwissenschaften (Dr.-Ing.)

genehmigte

Dissertation

von

Dipl.-Ing. Jens Kroneis
geb. in Kaiserslautern

D 386

Datum der mündlichen Prüfung	15. Dezember 2010
Dekan des Fachbereichs	Prof. Dipl.-Ing. Dr. Gerhard Fohler
Vorsitzender der Prüfungskommission	Prof. Dr.-Ing. Gerhard Huth
1. Berichterstatter	Prof. Dr.-Ing. Steven Liu
2. Berichterstatter	Prof. Dr.-Ing. Bernd Sauer

II

Acknowlegements

This thesis presents the results of my work at the Institute of Control Systems (LRS), Department of Electrical and Computer Engineering, at the University of Kaiserslautern. It is part of the project parallel kinematic structures, where our institute worked very successfully together with the Institute of Machine Elements, Gears and Transmissions (MEGT) at the University of Kaiserslautern.

I thank Prof. Dr.-Ing. Steven Liu for the excellent supervision of my work, the scientific discussions, the good working atmosphere at the institute and for his function as first referee of this thesis. Prof. Dr.-Ing. Bernd Sauer I would like to thank for inviting us to work together with his institute at the interesting research field of parallel robots, the possibility of using his demonstrators and the always very good cooperation with the members of his workgroup. Furthermore I would like to thank him for being the second referee of this thesis. Thanks are due to take the chair of this commission to Prof. Dr.-Ing. Gerhard Huth.

I would like to thank all my former colleagues at the LRS for the good time at the institute. For very inspiring and helpful technical discussions I would like to thank in particular Peter Müller, Nadine Stegmann, Christoph Prothmann and Daniel Görges.

Tobias Gastauer (MEGT) I would like to thank for our successful common work in the structural analysis and flexible modeling of parallel robots. Due to his help, that of Tim Leichner (MEGT) and Swen Becker the demonstrator has also always been operatable.

Furthermore I would like to thank all students working under my partly supervision on their student research projects and thesis as well as the IAESTE and RISE/DAAD trainees for their contributions and the interesting discussions. For supporting me during my time at the LRS I would like to thank all my student assistants: Mark Baumann, Patrick Demke, Stefan Hodek, Zhe Jiang, Peter Müller, Christina Strohrmann, Pedro Vallocci, Christoph Prothmann, Frank Rettinger and Jie Zhao.

My biggest thanks go to my parents and grandparents for their support and love. This thesis is dedicated to them.

IV

Contents

List of Figures

List of Tables

1 Introduction

In industrial practice most robots are of serial kinematic structure. They are highly flexible and cost-efficient. They have marked the industrial standard for many years. But a further improvement of their performance is, as a matter of the applied principle, limited and normally already outbid. In case of serial structures the inaccuracies and flexibilities of the single axes are accumulating [Abd07]. This disadvantage cannot be brought through. A higher stiffness typically results in a higher moving machine mass, reducing the achievable traveling (handling applications) or cycling (workpiece machining) times. For that reason the importance of parallel robots is increasing in time-sensitive, highly dynamical applications or processes requiring a high structural rigidity of the manipulator.

1.1 Motivation and aims of the thesis

Until today, most parallel robots have been operated by open-loop control concepts or closed-loop strategies applying strongly simplified dynamic models. The reason for the latter point is the high model complexity related to this type of manipulator structures. This aspect is further exacerbated in case of the in this thesis studied class of parallel kinematic robots containing functional or structural redundancies (named: complex parallel robots). In order to be able to apply sophisticated model-based control concepts for trajectory tracking anyhow directly to them, accurate, compact and real-time implementable dynamic models are necessary.

But compact dynamic models are only a first step on the way to a model-based trajectory tacking control with a high path accuracy and a low traveling time. Additionally, appropriately planned trajectories are required. For their planning it is necessary to consider the physical limits of the parallel robots, especially that of their drive systems. Also due to their several limbs this point is a challenging task, for which in case of parallel robots only first basic approaches are available. These two points are the key problems, which will be solved in this thesis, whose goal is the development of a holistic concept for the direct model-based trajectory tracking control of complex parallel manipulators.

The first key problem is solved in this thesis by introducing a new stepwise simplified modeling approach for complex parallel robots. It mainly bases upon the principle of equivalent lumped masses. The obtained simplified structures lead in combination with any standard dynamic modeling approach to accurate and compact models. These models can be directly (after the identification of their parameters) used to realize model-based control concepts like the *joint space inverse dynamics control*. If the dynamic models would be derived for the non-simplified structures (like normally) instead, this concept cannot be directly applied, due to its high computational load. In order to be able to apply it anyhow, it would be necessary to reduce the dynamic models, e.g. only considering its dominant terms. As a direct consequence the accuracy of the models and consequently that of the trajectory tracking would degrade.

In order to move the end-effector of a parallel robot time-optimally, but also vibration free along a predefined path a new strategy for the planning of path-constrained, smooth, time-optimal trajectories is proposed additionally. In contrast to other approaches all important constraints of the drive-systems are considered, while introducing a strategy appropriate for parallel robots. Both the model simplification and the trajectory planning strategy are general concepts, which can be applied to any (complex) parallel robot.

The solution of the two key problems is a main step on the way to the development of an overall concept for model-based trajectory tracking control of complex parallel robots. Further additional steps are necessary. In this context the velocity reconstruction can be mentioned exemplarily. Since the most model-based concepts for trajectory tracking control require information about all system states, which means especially that of the joint velocities, strategies for their determination have to be discussed. For solving all further subproblems occurring on the way to the model-based trajectory tracking control mainly state of the art strategies are applied in this thesis.

Finally, all parts come together for realizing the model-based trajectory tracking control in case of the SpiderMill, based on standard control algorithms. In this context also the interrelationships of the individual steps (chapters) are discussed in detail.

Summarized in one sentence: In this thesis a holistic concept for the model-based trajectory tracking control of complex parallel manipulators yielding a high tracking accuracy is presented and demonstrated on the 2-degrees of freedom (DOF) planar parallel robot SpiderMill, wherever necessary discussing its particularities.

1.2 Structure of the thesis

Each chapter of this thesis (except the second) is structured as a separate document, which means that each of it has its own introduction and conclusions. Due to this organization of the thesis, the work of others (state of the art – if not only standard approaches are applied) and maybe necessary further steps are discussed in a direct context to the performed analysis. Furthermore the chapters have been arranged from the view point of control.

In Chap. 2 general aspects of serial and parallel robots are discussed in a compact form and the demonstrator SpiderMill is introduced. A structural analysis of the planar parallel robot is performed in Chap. 3, which means in detail that its overall degree of freedom is analyzed, its direct and inverse kinematics are set up and a workspace and singularity analysis are performed. In Chap. 4 a new modeling strategy for parallel robots with redundancies is introduced, leading to a simplification of the manipulator structure before its equations of motion will be derived. Based on the simplified structure of the SpiderMill several dynamic modeling strategies are applied in Chap. 5 in order to derive and analyze its equations of motion. One of these models is picked out and used through the rest of the thesis in order to define some kind of red line. In Chap. 6 the rigid body and the friction parameters of the model are identified. A new strategy for the planning of path-constrained, smooth, time-optimal trajectories in case of parallel robots is introduced in Chap. 7 and reference trajectories for the SpiderMill are planned. In the style of *inverse dynamics control* the non-linear dynamic model of the SpiderMill is feedback-linearized (and stabilized) in Chap. 8. Already from the feedback linearization it becomes obvious that different system states have to be observed in order to be able to realize sophisticated control strategies. For that reason in Chap. 9 different observer structures are analyzed. Applying them different model- and non-model-based trajectory tracking control strategies are implemented for the SpiderMill in Chap. 10. The results of the thesis are summarized in Chap. 11, where also further general possible extensions of the work are shortly discussed.

In App. A all parameters of the SpiderMill are summarized. The fundamental concepts of dynamic modeling applied in this thesis are shortly recapitulated in App. B. In App. C commonly known basic strategies for trajectory planning are sketched. A nomenclature is given in App. D. In App. E a short English and in App. F a detailed German summary can be found. Publications and supervised student theses originated during the time of this work are listed in App. G. The vita of the author is given in App. H.

2 Parallel manipulators and the demonstrator SpiderMill

2.1 Serial and parallel manipulators

In the design phase of parallel robots various factors (e.g. the ratio of workspace to installation space) are optimized in order to use their inherent advantages in the best possible way with respect to a specific work task. The different criteria for the evaluation of a mechanical structure can be divided into two classes: global and pose-depending criteria. In order to avoid an excessive discussion of the different aspects, advantages and disadvantages of serial and parallel robots, a compact comparison can be found in Table 2.1, where the ratings have been made in a global sense. Certainly, in each class of robots special types have been designed in order to overcome some of these drawbacks, mostly at the expense of other decreasing factors. For the optimization special strategies exist. Realistically, the evaluation of an aspect can only be made with respect to a specific task. Generally desirable properties are written **bold** in Table 2.1 and aspects in this thesis discussed are shaded . The not further analyzed ones play a subordinate role for this work.

The advantage of the arrangement of the drives on the fixed base and therefore outside of the workspace has to be examined from different viewpoints. Although this arrangement improves the dynamic properties of the robots and expedites their lightweight construction, it causes a higher constructive and computational (with respect to the control system) effort [Ker94]. These 'small' disadvantages are relativized keeping in mind that in case of serial robots each drive has beside the mass of the handling object itself additionally to move that of all subsequent links and drives. In case of the SpiderMill a special indirect actuation principle, cp. Sec. 2.2, is used, reducing the moving machine masses.

Position errors, caused by joint and gear tolerances (backlash), are added up in case of serial robots, leading usually to non-negligible values and therefore enhanced requirements on the closed-loop position control. Furthermore, the effort for the transfer of energy and signals is much higher for serial robots.

Particularly remarkable is the aspect that parallel structures show, despite of their lower machine mass, good rigidity values and low inertia forces. Hence better

5

category	criterion		serial robot	parallel robot
geometry	workspace and installation space	workspace	large	small
		ratio	big	small
	complexity of the construction		low	high
	use of repeat parts		hardly possible	standard case
	material costs / mechanical effort		large	small
statics	equations	direct statics	demanding	simple
		inverse statics	simple	demanding
	structural rigidity		middle	large
	mass of the robot		large	small
	ratio object mass/robot mass		small	large
kinematics	equations	direct	simple	difficult
		inverse	difficult	simple
	mobility in the workspace		large	constricted
singularities	location		in workspace / at boundaries	in workspace / at boundaries
	classes		type I	type I-III
	DOF		lose	lose / gain
	closeness		minor	high
dynamics	equations	direct	complex	very complex
		inverse	complex	very complex
	movement/operational speed		in general slower	in general faster / very fast
	resistance against reaction forces		low	high / very high
	moved machine masses		relatively large	small / very small
	inertia forces		large	small
drives	separation of drives and workspace		no	(mostly) fulfilled
	required drive power		large	smaller
trajectory planning	path planning		normal	simple (cp. inverse kinematics)
	trajectory calculation		normal	very complex (under research)
	concepts		in general standardized strategies	complex (under research)
control concepts	real-time implementation	hardware requirements	middle	(very) high
		software requirements	middle	high
		multiprocessor system	not required	recommendable
errors	pose accuracy		accumulative	compensateable
	(pose-) repeat accuracy		low	high
	force error		compensateable	accumulative
	dimension faults of the structure		accumulative	compensateable

Table 2.1: Comparison of serial and parallel robots

machine dynamics can be expected from them in principle. The higher structural rigidity is also one of the reasons for a better positioning accuracy of this class of robots. Normally serial robots have a lower rigidity and also their ratio of object mass to robot mass is much lower than in case of parallel robots. One further advantage of parallel robots is the higher linear velocity of their end-effector compared to serial robots [Ker94].

2.2 The demonstrator SpiderMill

2.2.1 Introduction of the demonstrator

The demonstrator SpiderMill, shown in Fig. 2.1, is a prototype of a parallel kine-matic mechanism. The shear kinematic structure has been developed at the *Institute for Virtual Product Engineering* (formerly: *Lehrstuhl für Rechneranwendung in der Konstruktion*) and employed in a common project of the *Institute of Machine Elements, Gears and Transmissions (MEGT)* and the *Institute of Control Systems* at the University of Kaiserslautern. The planned area of application of

Figure 2.1: The demonstrator SpiderMill [KGLS07b]

the robot is the rapid prototyping of sample models. The term rapid prototyping denotes a multiplicity of techniques which have the fast manufacturing of design models as common goal. Based on these techniques and CAD data, it is possible to realize a prototype in an early design phase, which leads partly to a considerable reduction of planning phases. Conventional machine tools for prototype manufac-turing, which apply techniques like NC milling or turning and grinding, feature the handicap of large machine masses which have to be accelerated. This main disad-vantage of serial structures, cp. Sec. 2.1, has been considered in the design of the demonstrator. Therefore a parallel kinematic concept has been realized, leading to a comparatively light, highly dynamical robot constructed with as many repeat and standard parts as possible.

The above mentioned machining centers are clearly overdimensioned for the task due to the fact that almost all design models are made out of non-metallic materials. Therefore the SpiderMill has been designed for the milling (shape cutting) of soft basic materials e.g. block material, wood or model wax. That is to say that for the design of the mechanical structure as well as the dimensioning of the drive concept low process forces have been assumed. The constructive dimensioning as well as the realization of the SpiderMill have been done in [Cro00].

2.2.2 Design and geometric parameters

The SpiderMill comprises a double redundant closed-chain structure, constructed with only revolute joints and standard aluminum profiles. Rigid body simulation models of the robot based on a multibody systems (MBS) description have been implemented in MSC.ADAMS[1] (cp. Fig. 2.2) and SimMechanics (cp. Fig. 2.6). In addition to the physical demonstrator, they are used as reference models for the verification of the kinematic and dynamic modeling, for parameter identification and the development and test of control algorithms.

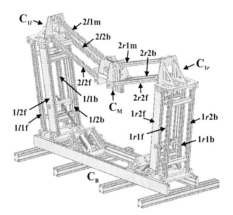

Figure 2.2: MSC.ADAMS model of the SpiderMill with structural analysis

---

[1]The model has been implemented at the Institute of Machine Elements, Gears and Transmissions at the University of Kaiserslautern.

In order to illustrate the symmetry of the SpiderMill indices for its left (l) and right (r) side with respect to the moving platform C_M are introduced. The index $k \in \{l, r\}$ is used if the side doesn't matter. Further indices are f and b for the front and back loop of the redundant structure, as well as m for links between the two loops. However, in almost all cases the parallel robot SpiderMill is considered as if it is planar, also neglecting the cross braces between the two redundant loops. The required kinematic parameters of the SpiderMill are introduced in Fig. 2.3 and Fig. 2.4.

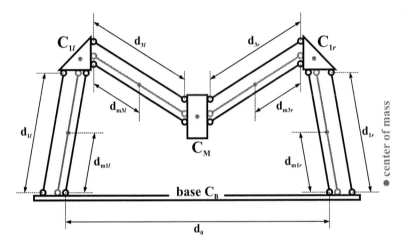

Figure 2.3: Kinematic parameters - part 1

Each of the two limbs of the SpiderMill consists of two redundant parallel crank mechanisms in series. According to [ES96] this type of gears has to feature a high production accuracy in order to avoid deadlocks. Therefore it is rarely used in newer mechanisms.

2.2.3 Joint variables and actuation principle

The spindle lengths s_k (distances $D_k A_k$ in Fig. 2.5) are defined as active prismatic joint variables $\mathbf{q}_a = [s_l, s_r]^T$. For obtaining a simple model the pseudo-active revolute joint variables $\mathbf{q}_{pa} = [\theta_l, \theta_r]^T$ and two passive joint variables $\mathbf{q}_p = [\gamma_l, \gamma_r]^T$

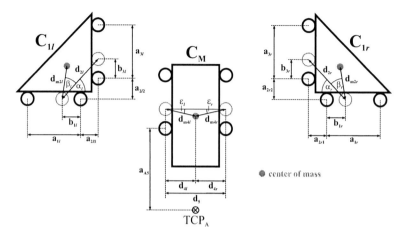

Figure 2.4: Kinematic parameters - part 2

are introduced additionally, see Fig. 2.5. The vector $\mathbf{q}_{pa}$ is called pseudo-active because its elements can be used as active joint coordinates to derive the inverse

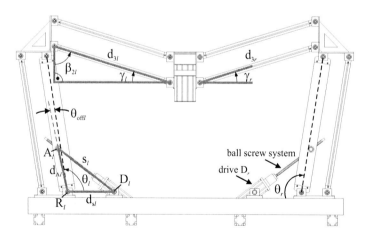

Figure 2.5: SpiderMill with geometric relationships and drive systems

kinematic and dynamic model, while disregarding the real actuation principle. Using the active joint variables $\mathbf{q}_a$, the vector $\mathbf{q}_{pa}$ itself can be considered as passive joint variables. The active and pseudo-active joint variables are related to each other by

$$s_k^2 = d_{Ak}^2 + d_{sk}^2 - 2d_{Ak}d_{sk}c(\theta_k + \theta_{offk}) \tag{2.1}$$

with $s(\alpha) = \sin(\alpha)$, $c(\alpha) = \cos(\alpha)$ and $t(\alpha) = \tan(\alpha)$. At the physical demonstrator the spindle bars are not mounted at the axis of symmetry of the actuated links. But it has been verified that assuming that the contact points A_k of the spindle forces f_{sk} (cp. Chap. 5) are lying there has a negligible influence on system dynamics within the workspace.

2.2.4 Experimental setup

The demonstrator SpiderMill has two drive systems, each consisting of a synchronous servo motor and a ball screw system. The drives can be operated in rotation speed control (maximal motor rotation speed: $3000 \frac{1}{\min}$) or like here done in torque control (nominal motor torque: 3.2 Nm). The reference values are given in form of voltage inputs to the servo controllers. As sensor systems two laser sensors mounted on the moving platform C_M (measuring range: 0.1 mm to 0.6 m) and motor encoders (number of increments: 1024) are available. The latter ones are used in this thesis due to their higher position resolution in combination with the ball screw systems to realize the trajectory tracking control algorithms in Chap. 10. In industrial applications laser sensors mounted on the moving platform are untypical. Here they are used for the initialization of the measurements, since the motor encoders only provide a relative information. Additionally, they define a rectangle by their measurement range, which is used as safety device by defining and monitoring a smaller square (0.05 m from each of its borders) within it. This area, called sensor-workspace in Sec. 3.3, is used in Sec. 7.3.1 for the path planning. The control structure is implemented on a dSPACE system (DS1104 R&D Controller Board), which can in case of the SpiderMill be operated with a sampling time of $T_S = 0.001$ s. The system processes the sensor data and controls the drives of the robot. Beside the control algorithms also the observers are implemented on it.

2.2.5 SimMechanics model of the SpiderMill

Especially in case of the development and verification of control algorithms the MSC.ADAMS model has several disadvantages, which are discussed in Sec. 5.6

in detail. In order to allow nevertheless an effective controller design a MBS model of the SpiderMill has been implemented using the SimMechanics Matlab toolbox [Mat07]. As in case of MSC.ADAMS the rigid bodies of the mechanism are implemented in form of concentrated masses. The theory behind SimMechanics is summarized in Sec. 5.6.1. With the SimMechanics model of the Spider-

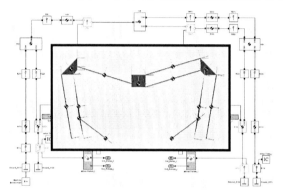

Figure 2.6: SimMechanics model of the SpiderMill

Mill, shown in Fig. 2.6, its direct and inverse dynamics can be simulated. Since the MSC.ADAMS and SimMechanics model show almost identical behavior, cp. Sec. 5.6.2, the latter can be used for the simulation and control of the robot. The motor torques $\boldsymbol{\tau}_k$ are defined as inputs of the SimMechanics model and the spindle length s_k and the end-effector position $^B\mathbf{x}_E$ as outputs.

Without requiring a co-simulation, sensor behavior and actuator dynamics can be implemented in SimMechanics in order to realize a more practical simulation. For the actuators the following identified transfer functions are implemented

$$G_{motor,l} = \frac{0.2823}{1.4376 \cdot 10^{-6} \cdot s^2 + 1.1487 \cdot 10^{-4} \cdot s + 1} \quad (2.2a)$$

$$G_{motor,r} = \frac{0.2834}{1.3983 \cdot 10^{-6} \cdot s^2 + 6.8275 \cdot 10^{-5} \cdot s + 1}. \quad (2.2b)$$

The two transfer-functions (2.2) describe the transfer behavior of the drives including the servo-controllers. In order to simulate the behavior of the encoders, the courses of the spindle length s_k are sampled with the sampling time T_S of the dSPACE system. Afterwards the sampled courses are quantized according to the number of encoder increments. Also the behavior of the laser sensors is simulated overlayed with noise. With this practical model the effects of the sensors and actuators on the control can be analyzed.

3 Structural analysis

The main problems in the context of parallel robots are their unfavorable ratio of working space to installation space and singularities within their workspace. Both the workspace and the singular configurations have to be analyzed before any motion can be planned. In the design phase of a robot these aspects are considered, while performing its structural optimization.

In this chapter the *mobility* of the planar parallel robot SpiderMill is studied based on an extended version of the (Chebychev-) Grübler-Kutzbach criterion considering passive DOFs and redundant constraints [HLL08]. In a further step the *kinematic equations* are derived applying a geometric approach [Tsa99] and also, very shortly summarized, the modified Denavit-Hartenberg (MDH) notation [KK86].

Based on the kinematic relationships the *workspace* of the robot can be analyzed. In the literature various methods for parallel robots have been proposed using geometric or numerical methods. A good overview about commonly applied strategies can be found in [Mer00]. The different types of workspaces, e.g. the constant orientation workspace, can be classified by the schemes given in [Tsa99, Mer00]. Especially the workspace boundaries and existing singularities play an important role in workspace analysis, limiting the end-effector motion. Beside the active also the passive joint limits have to be considered in order to determine the workspace of a manipulator correctly [PH05].

Most of the numerical methods for determining the workspace of parallel robots rely on the discretization of the pose parameter [CA91, FMS95]. The accuracy of the so determined workspace boundary depends on the sampling step size that has been used to create the grid [SMB07]. With decreasing step size the computation time grows exponentially. In this thesis the numerical inaccuracy at the boundary is eliminated and an analytically and geometrically interpretable description based on unwind circles determined. Even though nowadays powerful three-dimensional CAD packages and MBS programs are available, they cannot provide and visualize realistic workspace informations, which are essential in the manipulator design phase [SMB07].

One important aspect of the workspace analysis performed in this chapter is the fact, that all results obtained by studying the kinematic relationships have been confirmed applying standard approaches of the mechanisms and gear trains together with the MEGT.

Beside the information about the workspace boundaries also the knowledge about the existence of *singularities* within a connected workspace or singularities separating distributed workspaces from each other is important. The singular configurations within a connected workspace have to be addressed while planning the motion of the end-effector. The singularities connecting distributed workspaces can in several cases be overcome – e.g. by utilizing the force of gravity [BLH07, BKL+08] – considerably maximizing the usable workspace.

For the singularity analysis of parallel robots different methods exist. The classical ones [Tsa99, Mer00, GA90] use the conventional or screw-based Jacobian matrices, which means that they are based on the differential kinematics, in order to determine singularities. For singular configurations the matrices relating the input to the output velocities become rank deficient [GA90]. In a few cases the singularities can also be identified directly by analyzing the kinematic equations. In other cases they are determined based on a static analysis [Mer00], degenerated screws [Mer00], line geometry (in particular on Grassmann geometry [Mer00, MLC07]), graph theory [DW06] or distance dimensions [XKW92]. For all manipulators the singularity analysis is – like the derivation of the kinematic equations – an individual analysis step, which can only to some degree be generalized. This does also hold for the singularity analysis performed in this chapter for the SpiderMill. Its results are strongly limited to that individual manipulator. Nevertheless the information are very important for the path planning.

This chapter is structured as follows: A classification of the robot, mainly determining its DOFs, is performed in Sec. 3.1. By applying primarily a vector-loop analysis, its direct and inverse kinematics are determined in Sec. 3.2. The workspace and singularity analysis are performed in Sec. 3.3 and 3.4. Conclusions are drawn in Sec. 3.5.

3.1 Degrees of freedom of the SpiderMill

The degrees of freedom (DOFs) F of a mechanism are equal to the number of independent parameters needed to specify its configuration completely. Applying the standard (Chebychev-) Grübler-Kutzbach criterion in order to determine the DOFs of a mechanism with structural particularities regularly leads to a wrong result. Only when passive or identical DOFs [RAG+06] f_p and redundant constraints [Ker94, ES96] z are taken into account the DOFs can be determined correctly. Especially the last ones are difficult to identify and only a few approaches exist [DHL06]. In the following an extended version of the Grübler-Kutzbach criterion [HLL08], considering the particularities, is applied

$$F = \lambda \cdot (n - j - 1) + \left(\sum_{i=1}^{j} f_i \right) - f_p + z, \qquad (3.1)$$

with the number of links (including the fixed base) and joints (assumption: all joints are binary) given by n and j, f_i defining the DOFs of joint i and λ the DOFs of the space in which the mechanism is intended to work ($\lambda \in \{3, 6\}$).[1]

In order to determine the DOFs of SpiderMill correctly, its passive DOFs and redundant constraints have to be identified. By analyzing the DOFs with and without one of the crank levers, it becomes obvious, that one redundant constraint ($z = 1$) exists for the planarized model yielding a DOF of

$$F = 3 \cdot (12 - 16 - 1) + \left(\sum_{i=1}^{16} 1 \right) - 0 + 1 = 2$$

when Eq. (3.1) is applied. Due to the structural but not functional redundancy of the front and back loop this result also holds for the spatial demonstrator. Thus the SpiderMill is a planar, symmetrical, 2-DOFs, fully parallel robot and can be classified as: $2 - \underline{R}RRR$.

3.2 Direct and inverse kinematics

The kinematics of the simplified SpiderMill model will be determined based on the MDH method [KK86] and the geometric approach [Tsa99]. Whereas the latter one is presented in more detail, since it allows a further verification of the singular configurations in Sec. 3.4. In case of geometric approach, the following two vector-loop equations can be derived for the inner loop in Fig. 3.1

$$\overline{A_l F} = \overline{A_l B_l} + \overline{B_l C_l} + \overline{C_l D_l} + \overline{D_l E_l} + \overline{E_l F} \qquad (3.2a)$$

$$\overline{A_l F} = \overline{A_l A_r} + \overline{A_r B_r} + \overline{B_r C_r} + \overline{C_r D_r} + \overline{D_r E_r} + \overline{E_r F}. \qquad (3.2b)$$

Expressing the vector-loop equation (3.2a) in the fixed coordinate frame BI of the inner loop gives

$$x_{TCP,BI} = d_{1l} c\theta_l + a_{2l1} + d_{3l} c\gamma_l(t) + d_{4l} \qquad (= \Gamma_{1l}^*) \qquad (3.3a)$$

$$y_{TCP,BI} = d_{1l} s\theta_l + a_{2l2} - d_{3l} s\gamma_l(t) \qquad (= \Gamma_{2l}^*) \qquad (3.3b)$$

---

[1]In contrast to the classical criterion [Hun90, Wal66] ($\lambda = 3$: planar or spherical, $\lambda = 6$: spatial mechanisms), several newer approaches also adapt the value of λ in order to determine the correct DOFs of a mechanism [DHL06].

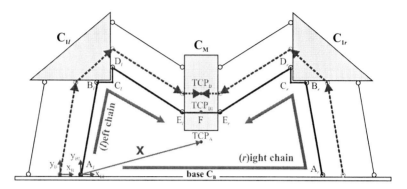

Figure 3.1: SpiderMill with vector-loops and end-effector vector $\mathbf{x}$

Eliminating the passive joint angle γ_l from Eq. (3.3) yields

$$
\begin{aligned}
n_1 = 0 =& (a_{2l1} + d_{4l} - x_{TCP,BI})^2 + (a_{2l2} - y_{TCP,BI})^2 + d_{1l}^2 - d_{3l}^2 \\
& + 2d_{1l}(a_{2l1} + d_{4l} - x_{TCP,BI})c\theta_l(t) + 2d_{1l}(a_{2l2} - y_{TCP,BI})s\theta_l(t).
\end{aligned}
\tag{3.4}
$$

Performing the same steps for the vector-loop equation (3.2b),

$$
x_{TCP,BI} = d_0 - d_{1r}c\theta_r(t) - a_{2r1} - d_{3r}c\gamma_r(t) - d_{4r} \qquad (= \Gamma_{1r}^*) \tag{3.5a}
$$

$$
y_{TCP,BI} = d_{1r}s\theta_r(t) + a_{2r2} - d_{3r}s\gamma_r(t) \qquad (= \Gamma_{2r}^*) \tag{3.5b}
$$

is obtained, finally leading to

$$
\begin{aligned}
n_2 = 0 =& (a_{2r1} - d_0 + d_{4r} + x_{TCP,BI})^2 + (a_{2r2} - y_{TCP,BI})^2 + d_{1r}^2 - d_{3r}^2 \\
& + 2d_{1r}(a_{2r1} - d_0 + d_{4r} + x_{TCP,BI})c\theta_r(t) + 2d_{1r}(a_{2r2} - y_{TCP,BI})s\theta_r(t).
\end{aligned}
\tag{3.6}
$$

Direct kinematic problem (DKP): The aim of the DKP is to determine the pose of the end-effector $\mathbf{x}$ as a function of the joint variables $\mathbf{q}$

$$
\mathbf{x} = \boldsymbol{\mathcal{D}}\left(\mathbf{q}\right), \tag{3.7}
$$

where the joint variables are the active $\mathbf{q}_a$ or like in the following the pseudo-active ones $\mathbf{q}_{pa}$. In order to solve the DKP Eqs. (3.4) and (3.6) are rewritten in the following form[2]

$$
x_{TCP,BI}^2 + y_{TCP,BI}^2 + f_{k1}x_{TCP,BI} + f_{k2}y_{TCP,BI} + f_{k3} = 0. \tag{3.8}
$$

---

[2]In f_{ki} further geometric relationships including the pseudo-active joint coordinates are summarized which are not presented here for a compact description.

Solving this system of equations yields

$$
\begin{aligned}
x_{TCP,BI_{1,2}} =& \frac{1}{2[(f_{l1} - f_{r1})^2 + (f_{l2} - f_{r2})^2]} \left(f_{r1}(-f_{l2}^2 + 2f_{l3} + f_{l2}f_{r2} - 2f_{r3}) \right. \\
&+ f_{l1}(-2f_{l3} + f_{r2}(f_{l2} - f_{r2}) + 2f_{r3}) + (\pm f_{r2} \mp f_{l2})\sqrt{rad1} \bigg) \quad (3.9a)
\end{aligned}
$$

$$
\begin{aligned}
y_{TCP,BI_{1,2}} =& \frac{1}{2[(f_{l1} - f_{r1})^2 + (f_{l2} - f_{r2})^2]} \left(f_{l2}(-f_{r1}^2 - 2f_{l3} + f_{l1}f_{r1} + 2f_{r3}) \right. \\
&+ f_{r2}(2f_{l3} + f_{l1}(f_{r1} - f_{l1}) - 2f_{r3}) + (\pm f_{l1} \mp f_{r1})\sqrt{rad1} \bigg) \quad (3.9b)
\end{aligned}
$$

with
$$
\begin{aligned}
rad1 =& -4f_{l3}^2 + (f_{l2}f_{r1} - f_{l1}f_{r2})^2 - 4f_{r3} \left(f_{l1}^2 - f_{l1}f_{r1} + f_{l2}(f_{l2} - f_{r2}) \right) \\
&- 4f_{r3}^2 + 4f_{l3}((f_{l1} - f_{r1})f_{r1} + (f_{l2} - f_{r2})f_{r2} + 2f_{r3}).
\end{aligned}
$$

Inverse kinematic problem (IKP): The aim of the IKP is to determine the posture of the joint variables $\mathbf{q}$ as a function of the end-effector pose $\mathbf{x}$.

$$
\mathbf{q} = \mathcal{I}(\mathbf{x}), \quad (3.10)
$$

where the joint variables are again in the following the pseudo-active ones. In order to calculate the pseudo-active joint angles Eqs. (3.4) and (3.6) are arranged in the following form[3]

$$
e_{k1} \cdot c\theta_k(t) + e_{k2} \cdot s\theta_k(t) + e_{k3} = 0 \quad (3.11)
$$

from which, following the steps proposed in [Tsa99], the IKP can be solved

$$
\theta_k = 2 \arctan \frac{-e_{k2} \pm \sqrt{e_{k1}^2 + e_{k2}^2 - e_{k3}^2}}{e_{k3} - e_{k1}} \quad (3.12)
$$

Yielding two solutions for each of the two active joint angles.

The so obtained direct and inverse kinematic descriptions are based on the inner vector-loop in Fig. 3.1. An equivalent model has also been derived for the dashed loop, where the complexity of the equations has been considerably reduced while improving the numerical accuracy (cp. Tab. 3.1) at the same time, by retaining the passive joint angles γ_k (as intermediate calculation step solving the DKP). In order to derive this second kinematic model the transformation matrices of the vector chains of the left (l) and right (r) side from the base frame B to the end-effector

---

[3]In e_{k_i} further geometric relationships including the TCP coordinates are summarized which are not presented here for a compact description.

frame TCP_B have been derived following the MDH notation. From the resulting transformation matrices two equations for the end-effector position are obtained, from which relationships of the following form

$$\gamma_l = u(\theta_l, \theta_r) \qquad \text{and} \qquad \gamma_r = v(\theta_l, \theta_r), \tag{3.13}$$

with u and v representing some nonlinear functions can be derived for $\mathbf{q}_p = [\gamma_l, \gamma_r]^T$. Due to their high complexity they are not presented here. Although, there are numerical advantages by calculating the passive joint angles at an intermediate step, the first derived kinematic description allows in case of a singularity analysis (cp. Sec. 3.4) a further verification based on the root terms.

In order to verify the direct and inverse kinematic descriptions the MSC.ADAMS model of the SpiderMill is used. For three reference trajectories (circle, triangle and eight) implemented in form of splines for s_l and s_r, the resulting pseudo active joint angles and TCP-coordinates have been measured in MSC.ADAMS. In order to perform a static analysis and verification of the kinematic descriptions for each point of the trajectories the measured joint angles have been used to calculate the TCP-coordinates (DKP) and the measured TCP-coordinates to determine the joint coordinates (IKP). In a further step the errors between the calculated ($calc$) and measured (MSC) values are examined and quantified by the average trajectory error

$$\Delta P = \frac{1}{N} \sum_{k=1}^{N} \sqrt{(x_{calc,k}(t) - x_{MSC,k}(t))^2 + (y_{calc,k}(t) - y_{MSC,k}(t))^2} \tag{3.14}$$

and the averaged angular error

$$\Delta A = \frac{1}{2N} \left(\sum_{k=1}^{N} (\theta_{l,calc,k} - \theta_{l,MSC,k}) + \sum_{k=1}^{N} (\theta_{r,calc,k} - \theta_{r,MSC,k}) \right) \tag{3.15}$$

for the N sampling points of the respective trajectory. For the three trajectories, defined within the sensor workspace of the SpiderMill, cp. Fig. 3.2, the values summarized in Table 3.1 are obtained for the two sets of kinematic equations (index 1: eliminated / index 2: not eliminated passive joint angles), confirming the analytical descriptions. From the results in Table 3.1 the better numerical performance of the second approach can be seen. It is mainly a result of the different composition of the trigonometric functions. All occurring errors are numerical ones: of the MBS program and Matlab.

---

[4]The circle trajectories defined in Chap. 7 describe two runs through the circle path. This is done in order to be able to easier identify configurations, where the controller gains are not high enough – leading to a drop down of the TCP on the first part of the trajectory – in the tuning process of the controllers in Chap. 10.

trajectory	N	ΔP_1 [m]	ΔA_1 [rad]	ΔP_2 [m]	ΔA_2 [rad]
circle[4]	633	$4.9240 \cdot 10^{-7}$	$5.2238 \cdot 10^{-8}$	$2.8884 \cdot 10^{-7}$	$2.3068 \cdot 10^{-8}$
triangle	450	$4.7111 \cdot 10^{-5}$	$1.0433 \cdot 10^{-5}$	$3.2169 \cdot 10^{-7}$	$1.7125 \cdot 10^{-7}$
eight	359	$1.6167 \cdot 10^{-5}$	$2.5546 \cdot 10^{-6}$	$3.4758 \cdot 10^{-7}$	$2.3245 \cdot 10^{-8}$
average	**480**	$\mathbf{2.2711 \cdot 10^{-5}}$	$\mathbf{4.8585 \cdot 10^{-6}}$	$\mathbf{3.1131 \cdot 10^{-7}}$	$\mathbf{8.6602 \cdot 10^{-8}}$

Table 3.1: Average trajectory and angular errors

3.3 Workspace analysis

Due to the fact that the parallel robot SpiderMill is a planar mechanism with an end-effector that can only change its position in a plane but not its orientation, only a reachable workspace has to be analyzed – which is here identical to the translational workspace [Tsa99, Mer00]. Moreover, the workspace of the robot can be divided into two parts – a lower utilizable and an upper theoretical one – which are connected by a contour of direct kinematic singularities (see Sec. 3.4 – K_{12} in Fig. 3.2). For practical applications only the lower workspace of the robot is of interest, cp. Sec. 3.3.5. It is a connected workspace without any singularities inside. Therefore it constitutes also a controllably dextrous workspace. For further aspects of workspace analysis cp. [Tsa99, Mer00]. To analyze the workspace of the SpiderMill and its boundaries two basic strategies can be applied. The first one is a purely geometric approach. It applies strategies of planar mechanisms and gear trains (in particular unwind circles) in combination with the limits introduced by contact situations due to the physical dimensions of the links and coupling elements as well as the limited lengths of the spindles. The other strategy is based on the discretization of the workspace by applying the kinematic equations of the robot and the constraints due to its closed loop structure. Of course also the contact situations and actuation limits have to be, additionally, taken into account for this approach – the latter, for a descriptive interpretation, expressed as limits of the pseudo-active joint angles $\mathbf{q}_{pa} = (\theta_l, \theta_r)$. The discretization based approach is presented in more detail below, also comparing its results with the purely geometric approach.

3.3.1 Limitation of the (pseudo) active joint variables

Besides the geometric parameters (lengths of the links), especially the minimum and maximum pseudo-active joint angles

$$\theta_{k,\{min,max\}} = -\theta_{offk} + \arccos \frac{d_{Ak}^2 + d_{sk}^2 - s_{k,min/max}^2}{2 d_{Ak} d_{sk}} = \left\{ \begin{array}{c} 30.3074° \\ 135.6448° \end{array} \right\} \quad (3.16)$$

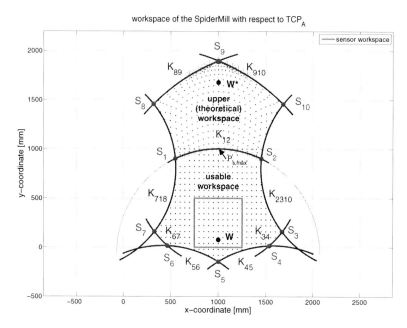

Figure 3.2: Workspace of the SpiderMill with boundary circles and sensor limits

restrict the workspace. However, due to the kinematic structure of the SpiderMill it is not possible that one of the pseudo-active joint angles reaches the maximum value and the other one the minimum at the same time.

The extreme values for equal (symmetric) joint angles are within the boundaries defined by Eq. (3.16). For maximum angles $\theta_{s,max}$, cp. Fig. 3.3 (a), the robot can be described by an isosceles trapezium, when the kinematic constants a_{2k1}, a_{2k2} and d_{4k} are relocated according to Fig. 3.3 (a). The corresponding joint coordinates can be calculated in two ways: planar geometry can be applied or they can be derived from Eq. (3.19a) (and (3.19b)) by assuming $\theta_{l,stretch} = \theta_{r,stretch} = \theta_{s,max}$ and $\gamma_{stretch} = 0°$, yielding

$$\theta_{s,max} = \arccos \frac{h_b - 2d_3}{2d_1} = 101.5369° \quad \text{with} \quad h_b = d_0 - 2a_{21} - 2d_4. \tag{3.17}$$

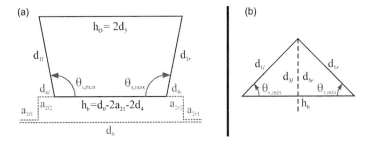

Figure 3.3: Maximum and minimum symmetric pseudo-active joint angles

In the following steps it becomes obvious that this case can occur and leads to a singular configuration (cp. Sec. 3.4). In Fig. 3.2 the related TCP-coordinates $TCP_{BI,A}$ mark the point $P_{s,max}$ on curve K_{12}. The theoretically possible minimal symmetric angle $\theta_{s,min}$ can also be analyzed based on a line model of the robot. For that case, shown in Fig. 3.3 (b), the links d_{3l} and d_{3r} coincide. Again two ways are possible for calculating $\theta_{s,min}$: one is to analyze the isosceles triangle, the other one is to derive the angle from Eq. (3.19a) (and (3.19b)) assuming $\theta_{l,stretch} = \theta_{r,stretch} = \theta_{s,min}$ and $\gamma_{stretch} = 90°$, yielding

$$\theta_{s,min} = \arccos \frac{h_b}{2d_1} = 44.6426° \qquad (3.18)$$

Due to the fact that the necessary conditions for γ_l and γ_r as well as the calculated value of $\theta_{s,min}$ can not be realized at the physical robot (discussed below) the obtained result for $\theta_{s,min}$ is of theoretical nature only.

3.3.2 Stretch out configuration

Due to the closed loop structure of the SpiderMill, its usable workspace is limited by several contour curves, cp. Fig. 3.2. One of this limit curves, K_{12}, plays a special role. It separates the lower usable workspace from the upper theoretical one. All points on this continuous curve mark stretch out configurations which are, in addition, kinematic singularities (see Sec. 3.4). For the case that the SpiderMill is in such a configuration and one of its (pseudo) active variables would be further enlarged, the kinematic chain would burst.

The stretch out configurations can be described very intuitively using the passive joint angles. By equating Eq. (3.3a) and (3.5a) as well as Eq. (3.3b) and (3.5b)

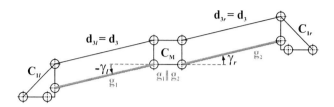

Figure 3.4: Stretch out configuration of the SpiderMill

the following set of loop-closure equations

$$0 = d_1 c\theta_l(t) + d_3 c\gamma_l(t) + d_1 c\theta_r(t) + d_3 c\gamma_r(t) - h_b \qquad (3.19a)$$
$$0 = d_1 s\theta_l(t) - d_3 s\gamma_l(t) - d_1 s\theta_r(t) + d_3 s\gamma_r(t) \qquad (3.19b)$$

is obtained for the closed kinematic chain of the SpiderMill.
For stretch out configurations (see Fig. 3.4) the two lines g_1 and g_2 (links d_{3l} and d_{3r}) are parallel to each other, making $-\gamma_l$ and γ_r to corresponding angles. Therefore a criterion for this configurations is

$$\gamma_{r,stretch} = -\gamma_{l,stretch} = \gamma_{stretch}. \qquad (3.20)$$

Applying this condition the closed kinematic chain can be described by

$$0 = d_1 c(\theta_{l,stretch}) + 2d_3 c(\gamma_{stretch}) + d_1 c(\theta_{r,stretch}) - h_b \qquad (3.21a)$$
$$0 = d_1 s(\theta_{l,stretch}) + 2d_3 s(\gamma_{stretch}) - d_1 s(\theta_{r,stretch}). \qquad (3.21b)$$

In Eq. (3.21) the passive joint angle $\gamma_{stretch}$ can be eliminated and one of the pseudo-active joint variables can be expressed with respect to the another. Due to their high complexity the resulting equations are not depicted. The results of the analysis have been verified using the CAD-tool AutoCAD comparing the points of the stretch out configuration with reference points obtained by the geometric approach.

3.3.3 Contact situations and interferences

Beside the limitations of the workspace due to the kinematic structure of the robot its constructional details are further restricting it. Therefore contact situations and interferences (in the sense of collisions of the limbs) will be analyzed in the following using AutoCAD. In Fig. 3.5 all possible contact situations in case of the SpiderMill are exemplarily depicted for the left side of the robot.

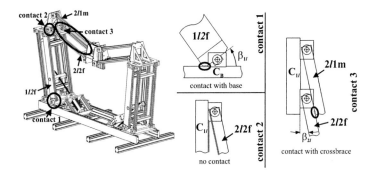

Figure 3.5: Contact situations for the SpiderMill

contact 1: A contact between the base C_B and the actuated link $1l2$ occurs for an angle of $\beta_{1l} = 36.3896°$. Due to the symmetry of the links both decreasing and increasing joint angles are limited,

$$53.6104° = (90° - \beta_{1l}) \le \theta_k \le (90° + \beta_{1l}) = 126.3896° \qquad (3.22)$$

contact 2: Between link $2l2$ and coupling element C_{1l} no contact occurs.
contact 3: It is possible that $2l1m$ bears on the cross braces between the front and the back loop. This situation can be expressed as a limit for angle $\beta_{2l} = 12.3092°$ between C_{1l} and $2l2$ or with respect to the passive joint angles

$$0° \le \gamma_k \le (90° - \beta_{2l}) = 77.6908°. \qquad (3.23)$$

In consideration of the restrictions introduced above, interferences of the limbs of the SpiderMill within its workspace can be precluded. This becomes obvious keeping the single loop structure and the parallel crank mechanisms of the planar parallel robot in mind.

3.3.4 Validity and limitations of the workspace analysis

When applying a discretization approach, especially the accuracy of the workspace boundaries, which constitute the essential information in case of the SpiderMill depends on the discretization step size. In contrast a purely geometric approach allows an exact construction of the limit curves using a CAD-tool. Together with the MEGT both approaches have been applied to the SpiderMill under the assumption, that the robot is described by a line model, whereas contact situations

and actuation limits are additionally taken into account. But in both cases the
geometric dimensions of the drive system and that of the moving platform have
been neglected.

In Fig. 3.2 the workspace of the SpiderMill is depicted with respect to the TCP
location of the MSC.ADAMS model ($TCP_A \rightarrow [A]$). It is possible to describe all
of its boundaries by circles, which are also sketched. Within the usable workspace
the sensor workspace is marked by a square. In this part of the workspace collisions
are impossible and no singularities occur.

A comparison of the discretization and the geometric approach shows identical
results for the usable workspace. In Fig. 3.2 the upper limit curve K_{12} as a result
of the geometric approach is marked by stars (∗) exemplarily. Each of its individual limiting/singular points has to be constructed separately using the purely
geometric approach.

3.3.5 The upper (theoretical) workspace

The upper workspace is simply the reflection of the lower one for allowable pseudo-
active and passive joint variables. From the singularity analysis in Sec. 3.4 it
becomes obvious that each set of allowed pseudo-active coordinates describes a
point in the lower (W) and in the upper (W^*) workspace, cp. Fig. 3.2. The
two resulting points are mirror images to each other with respect to the axis of
reflection defining a tetragon with d_{1l}, h_b and d_{1r} (cp. Fig. 3.7).

The upper and lower workspace of the SpiderMill are separated by the continuous
curve of singularities K_{12}. Due to the fact that the home or resting position of the
robot is within its lower (and therefore usable) workspace, a singular configuration
has to be crossed to get into the upper workspace. This aspect can be illustrated,
when the robot performs a symmetric motion in direction of increasing pseudo-
active joint angles, measuring the spindle forces in MSC.ADAMS. From Fig. 3.6 it

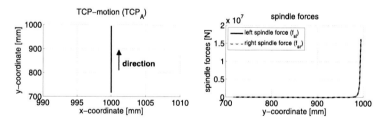

Figure 3.6: TCP-motion and related spindle forces for a symmetric motion

becomes obvious that the singular configurations (curve K_{12}, in Fig. 3.2) can not be overcome by the actuation of the robot.

3.4 Singularity analysis

3.4.1 Differential kinematics and Jacobian matrices

In case of the SpiderMill singular configurations cannot be identified directly based on the kinematic equations derived in Sec. 3.2. Therefore, in the following, the singularity analysis is performed based on the time derivatives of the loop-closure equations. In three steps, starting from the Jacobian matrices, inverse, direct and combined singularities are identified if existing, following the steps of analysis defined in [Tsa99]. Only structural possible configurations are considered and the case of constrained singularities is discussed.

The overall Jacobian matrix of a robot defines the relationship between its joint and end-effector velocities. This relationship is also called differential or velocity kinematics. The Jacobian matrix which is obtained by the direct differentiation of the kinematic equations with respect to time is called analytic Jacobian matrix [SK08]. In case of serial (sk) robots the matrix is defined the other way around than in case of parallel (pk) robots

$$\dot{\mathbf{x}} = \mathbf{J}_{sk} \cdot \dot{\mathbf{q}} \qquad \text{and} \qquad \dot{\mathbf{q}} = \mathbf{J}_{pk} \cdot \dot{\mathbf{x}}. \qquad (3.24)$$

In order to adapt the notation in case of the SpiderMill to that used in (3.24) and [Tsa99] the vector of the pseudo-active joint variables $\mathbf{q}_{pa}$ is denoted by $\mathbf{q}$ and the locations of the moving platform are described by a vector $\mathbf{x}$ (cp. Fig. 3.1) in the following. In general the kinematic constraints imposed by the limbs of a parallel robot can be described by an implicit function (cp. also Eq. (3.4) and (3.6))

$$\mathbf{n}\left(\mathbf{x}, \mathbf{q}\right) = \mathbf{0}. \qquad (3.25)$$

Its time derivative yields [Tsa99]

$$\mathbf{J}_x \cdot \dot{\mathbf{x}} = \mathbf{J}_q \cdot \dot{\mathbf{q}} \qquad (3.26)$$

$$\text{with} \quad \mathbf{J}_x = \frac{\partial \mathbf{n}}{\partial \mathbf{x}} \quad \text{and} \quad \mathbf{J}_q = -\frac{\partial \mathbf{n}}{\partial \mathbf{q}}.$$

The two obtained separate Jacobian matrices in Eq. (3.26) can be combined in the overall Jacobian matrices $\mathbf{J}_{sk}$ and $\mathbf{J}_{pk}$ according to Eq. (3.24), with

$$\mathbf{J}_{sk} = \mathbf{J}_x^{-1} \cdot \mathbf{J}_q \qquad \text{and} \qquad \mathbf{J}_{pk} = \mathbf{J}_{sk}^{-1} = \mathbf{J}_q^{-1} \cdot \mathbf{J}_x, \qquad (3.27)$$

whereas in case of parallel robots $\mathbf{J}_{pk} = \mathbf{J}$.

Deriving the partial derivatives of Eq. (3.4) and (3.6) the following two Jacobian matrices can be determined for the SpiderMill, assuming identical parameters for both sides of the robot and $\mathbf{q} = \mathbf{q}_{pa}$

$$\mathbf{J}_x = 2 \begin{bmatrix} x - d_1 c\theta_l - a_x & y - d_1 s\theta_l - a_{22} \\ x + d_1 c\theta_r - d_0 + a_x & y - d_1 s\theta_r - a_{22} \end{bmatrix} \tag{3.28a}$$

$$\mathbf{J}_q = 2d_1 \begin{bmatrix} (-a_{22} + y) c\theta_l + (a_x - x) s\theta_l & 0 \\ 0 & (-a_{22} + y) c\theta_r + (a_x - d_0 + x) s\theta_r \end{bmatrix} \tag{3.28b}$$

with $a_x = a_{21} + d_4$. Below, all three possible classes of singularities [Tsa99, GA90] will be analyzed for the SpiderMill. In order to allow a simpler geometric interpretation of the results, the kinematic constants a_{2k1}, a_{2k2}, d_{4k} are shifted towards its base, cp. Fig. 3.3 (a). This procedure has no effect on the obtained results and will be considered when TCP-coordinates are defined.

3.4.2 Inverse kinematic singularities (IKS) - Type 1

An IKS occurs when the determinant of $\mathbf{J}_q$ is equal to zero

$$\det(\mathbf{J}_q) = 0. \tag{3.29}$$

Interpretation: This type of singularity occurs when the closed kinematic loop reaches a workspace boundary (or an intern restriction). IKS separate the different solutions of the IKP, so called working modes [Bon02]. For this type of singularity different branches of the IKP meet – which is equivalent to the situation that the radian in Eq. (3.12) becomes zero. Typically this situation occurs, when a kinematic chain of the mechanism is either in a stretched out or folded back configuration [GA90]. For this type of singularity infinitesimal movements of the moving platform in certain directions are not possible. As a consequence the manipulator losses one or more DOFs [WH88].

Application to the SpiderMill: To find IKS the determinant of the Jacobian in Eq. (3.28b) has to be calculated and set to zero

$$\det(\mathbf{J}_q) = 4d_1^2 ((a_{22} - y)c\theta_l - (a_x - x)s\theta_l) ((a_{22} - y)c\theta_r - (a_x - d_0 + x)s\theta_r)$$
$$= 0. \tag{3.30}$$

Since Eq. (3.30) is difficult to interpret, the workspace boundaries will be analyzed in more detail to identify IKSs. Typical situations, occurring at the upper boundary (K_{89}, K_{910}, cp. Fig. 3.2) are sketched in Fig. 3.7. All configurations for which

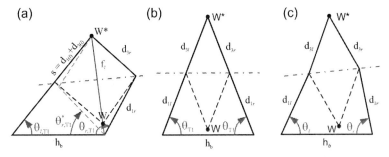

- · - · - · axis of reflection   – – – elbow configuration   ● point of lower (W) / upper (W*) workspace

Figure 3.7: Analysis of singularities of type 1

at least one limb of the robot is stretched out are describing points of the upper workspace boundary. This situation is exemplarily sketched for the left limb in Fig. 3.7 (a). For this case the following relationships between $\theta_{l,T1}$ and $\theta_{r,T1}$ (and $\theta_{r,T1}^*$ dashed case) hold

$$\theta_{r,T1} = \arcsin\left(\frac{s}{f_l} s\theta_{l,T1}\right) + \arccos\left(\frac{f_l^2 + d_1^2 - d_3^2}{2f_l d_1}\right) \tag{3.31a}$$

$$\theta_{r,T1}^* = \arcsin\left(\frac{s}{f_l} s\theta_{l,T1}\right) - \arccos\left(\frac{f_l^2 + d_1^2 - d_3^2}{2f_l d_1}\right) \tag{3.31b}$$

with $\qquad s = d_1 + d_3 \qquad$ and $\qquad f_l = \sqrt{h_b^2 + s^2 - 2\,h_b\,s\,c\theta_{l,T1}}$.

The related TCP-coordinates can be calculated as[5]

$$x_{TCP,BI,[A]} = s \cdot c\theta_{l,T1} + a_x \tag{3.32a}$$

$$y_{TCP,BI,[A]} = s \cdot s\theta_{l,T1} + a_{22} - [a_A]. \tag{3.32b}$$

A similar relationship can be derived for a stretched right limb.

For all configurations for which at least the links d_{1k} and d_{3k} of one side of the robot are on a line, the Jacobian matrix in Eq. (3.30) becomes equal to zero and the radian in Eq. (3.12) vanishes. This also holds for the special case of two stretched links, shown in Fig. 3.7 (b). For that case the line model of the robot is

---

[5]The notation [A] means, that in case of the MSC.ADAMS location of the TCP the offset a_A has to be considered.

describing an isosceles triangle, allowing to calculate the corresponding joint angles by applying the cosine rule

$$\theta_{T1} = \arccos\left(\frac{h_b}{2s}\right) = 68.1473°. \tag{3.33}$$

The corresponding TCP-coordinates are that of $S_{9[A]}$ in Fig. 3.2.

Due to the fact that the feasible joint angles are limited (cp. Sec. 3.3) only parts of the upper workspace boundary are described by IKSs. These continuous parts of K_{89} and K_{910} are marked by ∗ in Fig. 3.2. The remaining parts are described by configurations like that in Fig. 3.7 (c) for which the two upper links d_{3k} face upwards without any stretched limb. These parts of the limit curves are no restrictions in the sense of stretched out configurations but those imposed by contact situations. The TCP-coordinates for that configurations are

$$x_{TCP,BI,[A]} = d_1 c\theta_l + d_3 c \left[\arcsin\left[\frac{d_1(s\theta_r - s\theta_l)}{\sqrt{rad2}}\right] \right. \tag{3.34a}$$

$$\left. + 2\arctan\left[\sqrt{\frac{2d_3 - \sqrt{rad2}}{2d_3 + \sqrt{rad2}}}\right]\right] + a_x$$

$$y_{TCP,BI,[A]} = d_1 s\theta_l + d_3 s \left[\arcsin\left[\frac{d_1(s\theta_r - s\theta_l)}{\sqrt{rad2}}\right] \right. \tag{3.34b}$$

$$\left. + 2\arctan\left[\sqrt{\frac{2d_3 - \sqrt{rad2}}{2d_3 + \sqrt{rad2}}}\right]\right] + a_{22} - [a_A]$$

with $\quad rad2 = h_b^2 + 2d_1(d_1(1 + c(\theta_l + \theta_r)) - h_b(c\theta_l + c\theta_r))$.

They are parts of the continuous workspace boundary but they are no singularities. These results correspond to that of [BZG03], where it is shown that workspace boundaries are not per definition singularities of the first type.

Result: All configurations for which at least two links d_{1k} and d_{3k} are on one line are IKSs (type 1). All other configurations describing parts of the upper workspace boundary (K_{89}, K_{910}) are not singular.

3.4.3 Direct kinematic singularities (DKS) - Type 2

A DKS occurs, when the determinant of $\mathbf{J}_x$ is zero

$$\det(\mathbf{J}_x) = 0. \tag{3.35}$$

Interpretation: In contrast to the IKSs the DKSs are always inside the workspace of a parallel robot [GA90] and therefore more difficult to identify. In general, the different solutions of the DKP, called assembly modes [Bon02], are separated by DKSs for which the different branches of the solution of the DKP coincide. In case of DKSs the moving platform gains DOFs [WH88]. This means, it can carry out infinitesimal motions in some directions, while all actuators are locked. Or in other words, the robot cannot resist generalized forces in some directions [Tsa99].

Application to the SpiderMill: In order to determine DKSs the determinant of Eq. (3.28a) has to be calculated and set to zero

$$
\begin{aligned}
\det(\mathbf{J}_x) =& 4 \cdot ((2a_{21} - d_0 + 2d_4)(a_{22} - y) + d_1((a_{22} - y)(c\theta_l + c\theta_r) \\
& + (a_{21} - d_0 + d_4 + x)s\theta_l + (a_{21} + d_4 - x)s\theta_r + d_1 s\theta_{lr})) \\
=& 0
\end{aligned}
\tag{3.36}
$$

with $s\theta_{lr} = \sin(\theta_l + \theta_r)$. A direct analysis of Eq. (3.36) in order to find DKSs is not possible due to its complexity. Thus a more sophisticated approach is necessary. From the workspace analysis in Sec. 3.3 the special role of the limit curve K_{12}, defining so called stretch out configurations, becomes obvious. Thus it is self-evident to analyze these special configurations – separating the lower from the upper workspace of the SpiderMill – in the context of DKSs, typically occurring within the workspace of parallel robots. In order to do so, two further steps are possible: one is to eliminate $\gamma_{stretch}$ from Eq. (3.21) and to express one of the pseudo-active joint variables with respect to the other, yielding a very complex expression. The other one, yielding a compact formulation by applying plane geometry, is sketched below.

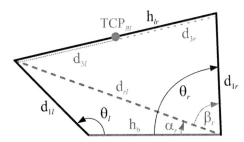

Figure 3.8: Kinematic structure of the SpiderMill with quadrangle

For stretched out configurations like that in Fig. 3.8, the four sides d_{1l}, h_{lr}, d_{1r} and h_b are defining a general quadrangle with the diagonal d_{rl}. The parameter h_{lr}

connecting the two halves of the robot is constant for stretch out configurations and is equal to $d_{3l} + d_{3r}$. Applying plane geometry θ_r can be expressed as a function of θ_l and known kinematic parameters

$$\theta_r = \alpha_r + \beta_r = \arcsin\left(\frac{d_1}{d_{rl}}s\theta_l\right) + \arccos\left(\frac{d_1^2 + d_{rl}^2 - h_{lr}}{2d_1 d_{rl}}\right) \tag{3.37}$$

with $\quad h_{lr} = d_{3l} + d_{3r} = 2d_3 \quad$ and $\quad d_{rl} = \sqrt{d_1^2 + h_b^2 - 2d_1 h_b c\theta_l} \; .$

The corresponding TCP-coordinates can be calculated as

$$x_{TCP,BI,[A]} = \frac{1}{2}\left(h_b + d_1 c\theta_l - d_1 c\theta_r\right) + a_x \tag{3.38a}$$

$$y_{TCP,BI,[A]} = \frac{1}{2} \cdot d_1 \left(s\theta_l + s\theta_r\right) + a_{22} - [a_A] \; . \tag{3.38b}$$

The calculation of the Jacobian matrix in Eq. (3.36) for all possible stretched out configurations (computed values are varying between $-4.44 \cdot 10^{-15}$ and $4.44 \cdot 10^{-15}$ for 73 001 points on the circular arc between S_1 and S_2, cp. Fig. 3.2) confirms the assumption that K_{12} is a continuous curve of inverse kinematic singularities. Based on this knowledge, the direct kinematics has been analyzed for the corresponding pseudo-active joint angles leading to the result that its two different solutions, cp. Eq. (3.9), always coincide for these configurations. It should additionally be mentioned that the results of the Jacobian analysis are even more important due to the fact that in case of some planar parallel robots, it is also possible to switch from one solution set of the DKP to another without encountering a DKS [Tsa99, IPC92, HM98, Cha98].

Result: All (stretched out) configurations for which the two links d_{3l} and d_{3r} of the robot are parallel to each other (curve K_{12}) are DKSs (type 2).

3.4.4 Combined singularities (CS) - Type 3

A CS occurs when both determinants (3.30) and (3.36) are zero at the same time

$$\det\left(\mathbf{J}_q\right) = 0 \qquad\qquad \text{and} \qquad\qquad \det\left(\mathbf{J}_x\right) = 0. \tag{3.39}$$

In contrast to IKSs or DKSs it can only occur for parallel robots with a special kinematic structure or architecture, respectively. The are not only configuration dependent. Not only the posture of the limbs, but rather the ratio of the lengths of the links among each other or with respect to the dimensions of the moving platform [Tsa99, GA90] are responsible whether a CS occurs or not. For a CS Eq. (3.25) degenerates [Tsa99, GA90, Rus88]. The moving platform can undergo some infinitesimal movements in some directions while all actuators are locked. On the other hand the platform can remain in a stationary pose while the actuators perform some infinitesimal movements.

Application to the SpiderMill: It is structurally impossible that the links d_{1k} and d_{3k} of the SpiderMill are in one line while its links d_{3l} and d_{3r} are parallel to each other at the same time.

Result: Singularities of the third type can be excluded for the SpiderMill.
Due to the actuation principle of the SpiderMill and the existent contact situations only the region above the base can be regarded as usable workspace of the robot. Therefore a singularity analysis for the other direction is not necessary, since no workspace enlargement in that direction is possible either.

3.4.5 Constrained singularities (CTS)

Another class of singularities, which cannot be identified based on the input-output relationship defined in Eq. (3.26) – which is also not valid for this case – are the so-called constraint singularities (CTS). This type of singularity occurs in case of parallel robots with reduced freedoms, when the screw systems formed by the constrained wrenches in all legs loses rank [ZBG02]. In case of classical planar and spherical parallel mechanisms all legs have the same DOFs as the whole mechanisms. In case of constrained parallel robots the individual leg chains impose different constraints. Only their combination yields the desired DOFs of the end-effector. Since the SpiderMill is not a constrained parallel robot and CTSs only occur for them, their existence in case of the SpiderMill can be excluded.

3.5 Conclusions and future works

In this chapter a structural analysis has been performed for the planar parallel robot SpiderMill, first determining its DOFs and classifying it. Mainly applying a geometric approach, the DKP and IKP has been solved. Based on the kinematics the workspace of the robot has been analyzed considering active and passive joint limits as well as contact situations and interferences. By analytically and therefore exactly describing the inner and outer workspace boundaries by limit circles, the main disadvantage of classical numerical approaches has been overcome. In a further step a complete singularity analysis has been performed for the SpiderMill based on the differential kinematics of the robot. In doing so, occurring singularities have been geometrically interpreted.

The results of the workspace and singularity analysis form the basis for forthcoming path and trajectory planning strategies. Moreover, open-loop control strategies can directly be realized based on the kinematics.

4 Simplified modeling approach

Different approaches are known for achieving a simplified model of a manipulator. Either the model is derived for the real physical robot and later reduced using a standard model reduction strategy (e.g. balanced truncation method [CLVD06, ZDG96]) or the structure of the robot itself is simplified (with reduced number of bodies) before deriving its equations of motions. Here a new strategy for the last approach is presented[1]. It is largely based on the well known concept of equivalent point or lumped masses [Diz66, Hoc70] which is used here in several ways for achieving a structural simplification. The strategy leads, in combination with any standard dynamic modeling approach (cp. Chap. 5), to a closed form and compact parameter-linear dynamic model. This is in particular relevant in case of complex parallel robots. Whereas the parameter linearity of the model is important in the context of parameter identification (cp. Chap. 6), allowing the application of linear estimators, its compactness is essential for the real-time implementation of sophisticated model-based control strategies (cp. Chap. 10).

The basis principle has also been applied by others, but without obtaining accurate, closed form, compact and parameter-linear models: In [Sta97] the equivalent point principle is applied to parallel robots with simple limb structure. Low weighted connection rods are presented by halved masses added to adjacent joints, assuming that the effect on full system dynamics is negligible. The separate description of the moving platform and the legs proposed in [LLL93] does not result in a closed form of dynamic description and thus does not allow the usage of linear identification strategies (cp. Chap. 6). In some cases [PKB+05] the models remain very complex even after the concept of lumped mass has been applied. Therefore a new more suitable strategy is proposed below.

This chapter is structured as follows: In Sec. 4.1 the new strategy (for parallel robots) is introduced, which is in Sec. 4.2 applied to the SpiderMill. Conclusions are drawn in Sec. 4.3.

---

[1]The in the following presented approach is taken (cited) from [KML09]. In doing so detected faults of the publication have been corrected.

33

4.1 Simplified modeling of parallel robots

Applying a standard modeling approach for deriving the dynamic equations of a robot, the complexity increases in cases of robots with redundant links and redundant passive joints. The goal of the here proposed simplified modeling strategy is to overcome these redundancies before deriving the equations of motion and to retain the kinematic and dynamic properties of the physical robot. To accomplish that, in the following possible cases of redundancies are defined and a stepwise strategy for simplified modeling, applying the concept of equivalent lumped masses, is described.

4.1.1 Concept of equivalent lumped masses

In general, the concept of equivalent lumped masses can be applied on a distributed mass, a group of interconnected masses or even a system of particles. It can be applied to replace rigid bodies by discrete point masses, without changing the dynamic behavior of the modified system [Hoc70, PK00].

In order to assure that the original and the equivalent structure are describing the same dynamic behavior the following conditions have to be met:

- their total mass m is equal

- they have the same center of mass $\mathcal{G}$

- they have the same moment of inertia $I_{\mathcal{G}}$ about the centroid axis.

These three conditions are described by [Diz66]

$$\sum_{i=1}^{n} m_i = m, \qquad \sum_{i=1}^{n} m_i \cdot \mathbf{x}_i = 0, \qquad \sum_{i=1}^{n} m_i \cdot l_i^2 = I_{\mathcal{G}}, \qquad (4.1)$$

where l_i is the distance between the position $\mathbf{x}_i = [x_i, y_i]^T$ of the point mass m_i and the position of center of gravity of the total mass m with respect to a coordinate system in[2] $\mathcal{G}$. $I_{\mathcal{G}}$ is the moment of inertia and n the number of individual masses. For the case of a planar structure, exemplarily shown in Fig. 4.1 for a system of three masses, $l_i = \sqrt{x_i^2 + y_i^2}$.

If only the first two conditions (4.1) hold, the substitution is called statical replacement [JH02]. If also the third condition is met, the mass m is called lumped mass, leading to a dynamic replacement. The application of the principle to a spatial structure can e.g. be found in [WG07], where it is used for examining the possibilities of replacing moving platforms by dynamically equivalent point masses.

---

[2]For a different location of the reference coordinate system the second and the third condition in Eq. (4.1) change.

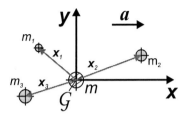

Figure 4.1: System with three point masses

4.1.2 Classes of suitable parallel robots

In the following different classes of parallel robots, suitable for applying the below presented modeling strategy, based on the lumped masses, are introduced and discussed. All of them are characterized structurally by some kind of redundancy (redundant links and passive joints).

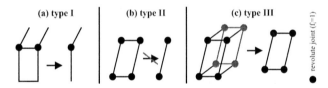

Figure 4.2: Types of redundancies

Type I redundancy: For the first group of parallel robots additional links and joints are added within the limbs, which do not contribute to the DOF of it and the overall mechanisms [Cod98,LGZ97,PDF91]. The redundancies can be replaced by single links and joints, cp. Fig. 4.2 (a), without changing the kinematics. One motivation for this kind of redundancy in mechanism design is the higher structural rigidity of the manipulator.

Type II redundancy: Here the redundancy has also an effect on the DOFs and therefore on the kinematics of the robot [HS92]. A good example for this form of redundancy are parallel crank mechanisms in a limb. They cannot be replaced by single links and joints, cp. Fig. 4.2 (b), if further joints and links are connected to them.

Type III redundancy: A combination of both forms of redundancies is possible [HCMZ06,PKB$^+$05]. The type I redundancy can be a part of the group of links describing a type II redundancy (e.g. compare Fig. 4.7) or two identical structural elements (planar gear mechanisms, e.g. parallel crank mechanisms) perform the same movements in two parallel planes, cp. Fig. 4.2 (c). In this case the individual links of the functional elements are often connected by cross braces. These braces enhance additionally the structural rigidity of the robot, but do not change the functionality or the kinematics of the group of links. It should be mentioned that a combination of two or more type II redundancies is also a type III case.

Redundancies, especially that of the type II and their combinations, are very common in parallel robots. As stated in [LW03], planar four-bar parallelograms are frequently applied as construction elements in the design of parallel robots to improve their rotational capability and their stiffness. In all three cases the redundancies lead to a higher modeling effort regardless of the used approach.

4.1.3 Strategy for simplified modeling

The proposed modeling strategy consists of the following steps.

Step 1 - structural analysis: The parallel robot is analyzed and coupling elements C_i, the moving platform C_M and the base C_B of the robot – the last two are both special cases of coupling elements – are identified. In the following coupling elements will be understood as elements connecting groups G_i of links L_i with redundant character, cp. Fig. 4.7. Furthermore, the functionality of the robot as a whole and that of its individual groups G_i is examined by considering the types of redundancies in the mechanism.

Step 2 - substitution of redundancies: Groups of links G_i with redundancy – which are always separated by coupling elements – are substituted by point masses m_i by applying the concept of equivalent lumped masses. As a consequence of this step the number of bodies N_k in the system decreases and the dynamic properties are kept. In case of type I redundancies also the kinematic properties are preserved and the system remains statically determinate, whereas type II and III redundancies need further consideration (see step 4). In all cases substitution parts S_i with line geometry are introduced. They have the same orientation (and length) as the original links of the group and pass the location of the point masses m_i, cp. Fig. 4.3.

Also with respect to parameter identification redundancies become important. Groups G_i of elements (links L_i) whose parameters cannot be identified separately (e.g. L_{Gi1} and L_{Gi2} in Fig. 4.3 (b)) have to be handled as one part S_i.

Step 3 - dealing with coupling elements: The coupling elements C_i and the moving platform C_M are substituted by point masses m_{Ci} and m_{CM} which

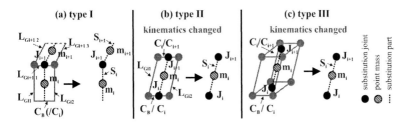

Figure 4.3: Substitution of redundancies

are located in the center of gravity of the respective elements. Similar to other approaches [PBBH06, PK00] the masses of the joints connecting these elements with the adjacent links are also combined in the respective point masses.

Step 3 (a) - joints of the simplified model: To replace the original joints between two coupling elements by substitution joints J_i, cp. Fig. 4.3, the type of redundancy has to be considered. If the corresponding groups are of type I redundancy, cp. Fig. 4.3 (a), the original joints can be directly replaced by single joints at both ends of S_i. The substitution joints have the same DOFs and are of the same type as the original joints. In case of type II or III redundancies the ends of S_i are also chosen as locations for J_i, see Fig. 4.3 (b) and (c), but their DOFs or even the kinematic structure of the compact model itself has to be adapted to preserve the functionality of the original groups (see step 4). For planar parallel robots the original joints and the substitution joints are only elementary prismatic or rotatory joints.

Step 3 (b) - point mass(es) for the coupling elements: So far the coupling elements are represented by the point masses m_{Ci} and m_{CM}. Their further treatment depends on the geometry of the physical body, which is used as a coupling element, and on the locations of the substitution bodies and joints relative to the positions of the point masses. In the following some possible configurations are discussed.

- If the distances δ_1 and δ_2 between the J_i and the location of the point mass are relatively small compared to the overall kinematic dimensions[3], e.g. $\delta_1 \ll 1$, $\delta_2 \ll 1$ in Fig. 4.4 (a), the coupling element C_i should further on be represented by a single point mass m_{Ci} located in its center of mass. Furthermore, for very small δ_i, it is in many cases appropriate to replace the whole element with its belonging J_is by a single substitution joint J_{mi} and locate the point

---

[3]For a compact representation of the situation of small distances compared to the overall dimensions of a rigid body the notation $\delta_i \ll 1$ is used in this chapter.

mass in it, cp. Fig. 4.4 (b).

- Another possibility to treat the coupling elements is to separate the single point masses representing C_i or C_M in several equivalent masses located in the substitution joints J_{mi}, see Fig. 4.4 (c). This approach, also used in [GVCW04, WG07] for a different purpose, is also an application of the principle of equivalent masses. The concrete way of the division of the single point mass thereby depends on the shape of the coupling element, the locations of J_is relative to each other (e.g. if they are in line) and the operational space of the robot (planar or spatial). Depending on the concrete configuration it would be in some cases also necessary to change the locations of the joints relative to each other in order to fulfill the equivalence requirements.

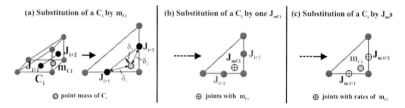

Figure 4.4: Substitution of coupling elements

In general, the treatment for both types of coupling elements C_i or C_M is the same. However, special considerations can be reasonable in case of very high payloads or strongly unsymmetrically mounted masses on the moving platform C_M, see Fig. 4.5. Whether the payload should be integrated in m_{CM} or directly represented by a separate point mass m_{load}, is to be decided case by case. Especially in the case that the assumption $\delta_3 \ll 1$ is no more valid, the modeling of the load as an individual point mass is more reasonable than to combine it in m_{CM}.

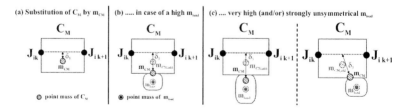

Figure 4.5: Substitution of C_M generally and in case of high payloads

Step 4 - statically determinate model - retaining of kinematics: As mentioned in step 2 and 3 additional considerations are needed for replacing type II and III redundancies to achieve the kinematic equivalence of the resulting simplified structure. It is always possible to transform a type III redundancy to a type II one by reducing the functional redundancies in different planes to the virtual plane E_V where the centers of masses of the involved groups are jointly located, e.g. compare Sec. 4.2. One possible, though not very intuitive way to handle the type II redundancies is to kinematically equivalently shift all coupling elements. An example of such a treatment is given in Sec. 4.2. Nevertheless, there also exist some more systematic approaches for obtaining a statically determinate model with unchanged kinematics. These are usually design methods and mostly applied for the synthesis of lower-mobility parallel mechanisms [FCSB00, HS91, HL03, JYAXLL01, KH02, KG02, LHH04].

Step 5 - modeling of actuation: In case of direct driven joints the actuation takes place via J_i or J_{mi}, respectively. In this case, especially the considerations of step 3 (a) and 4 apply. For an indirect actuation of the joints, however, additional considerations may be useful to find proper 'active' or 'pseudo-active' joints (e.g. the input selection method [Hua04, HL03, ZH00, ZH07]). An appropriate choice can further simplify the structural analysis.

Since most parallel robots contain redundancies, the application of the proposed strategy can in some cases dramatically reduce the modeling effort.

4.1.4 Modeling errors

In this section some considerations regarding the modeling errors are made. These errors are mainly connected with the handling of the coupling elements and those in combination with the simplification of type II or III redundancies, often caused by the required adaption of robot kinematics and dynamics. In the following different cases are considered.

Situation 1 (a): If a type II or III redundancy exists between two coupling elements with one of which is the base C_B, the simplification by applying the principle of equivalent lumped masses and shifting the mass of the coupling element m_{Ci} to the substitution joint J_{mCi} (assuming $\delta_1 = 0$) does not introduce a modeling error. An example for such a configuration and its simplification is shown in Fig. 4.6. The dynamic models for the original structure and its simplification are identical for

$$m_i = m_{LGi1} + m_{LGi2}, \qquad (4.2a)$$
$$I_i = I_{LGi1} + I_{LGi2}. \qquad (4.2b)$$

The same result is obtained using equivalent lumped masses. The constant δ_1 has no influence on the system dynamics. Therefore no modeling error occurs. An evidence for this statement is given in Sec. 5.2.1 where the situation is analyzed in case of the SpiderMill in detail.

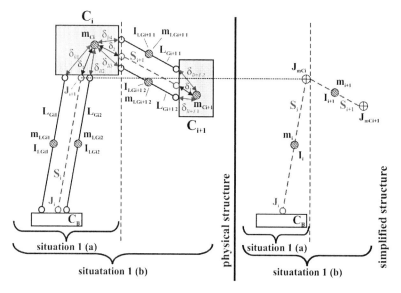

Figure 4.6: Evaluation of the modeling error exemplarily for planar four-bar parallelogram structures

Situation 1 (b): Assuming a mechanical structure where in contrast to situation 1 (a) the coupling element C_i is additionally connected to a subsequent coupling element C_{i+1} by a group of links, see Fig. 4.6, the application of the proposed simplification strategy introduces a modeling error. But this error is not caused by the disregard of the lever arms ($\delta_{li} = 0$). The results of a concrete analysis in case of the SpiderMill in Sec. 5.2.2 show, that they are not part of the analytical model derived, without performing the simplification strategy. Also deriving the dynamic equations of the whole SpiderMill in Sec. 5.1.3 they do not get a part of the solution. Based on these results it is assumed, that the introduced error is a consequence of the different possible motions of the physical and the simplified structure.

Situation 2: If the point masses m_{Ci} or m_{CM} representing a coupling element are divided and added to the substitution joints (see Fig. 4.4 (c)) the size of the individual δ_i is directly correlated to the modeling error. For example, in case of a coupling element whose two adjacent substitution joints are lying in one line with the point mass m_{Ci}, which implies $\delta_3 = 0$ (see Fig. 4.4), no modeling error will be introduced when applying the principle of equivalent lumped masses [WG07]. If, however, the three elements are not in a line, $\delta_3 \neq 0$, a modeling error will occur when step 3 is performed and the joint locations are kept. This error increases for larger values of δ_3.

4.2 Simplified modeling of the SpiderMill

The modeling strategy proposed in Sec. 4.1.3 is in the following exemplarily applied to the SpiderMill for finding a simplified structure retaining the kinematic and dynamic properties of the physical robot.[4]

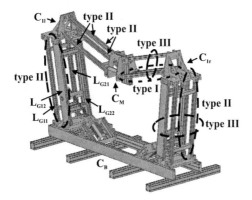

Figure 4.7: MSC.ADAMS model of SpiderMill with structural analysis

Step 1 - structural analysis: The planar parallel robot SpiderMill, cp. Fig. 4.7, consists of a base C_B, a moving platform C_M and two more coupling elements C_{1l} and C_{1r}. It has two translational DOFs and each limb consists of three parallel crank mechanisms in form of two type III redundancies. The first one is between

---

[4]In Sec. 5.1.1 additionally a further simplified structure – obtained by not performing all steps of the simplification strategy like described below – is introduced. Based on it the influence of some of the simplification steps can be further analyzed.

the coupling elements C_{1k} and C_B (e.g. links L_{G11}, L_{G12} and L_{G21}, L_{G22} – a combination of two type II redundancies) and the other one between C_{1k} and C_M (a combination of one type I redundancy and some kind of type II redundancy).

Step 2 - substitution of redundancies: Substituting the redundancies the simplified structure in Fig. 4.8 is obtained. The positions of the S_is are derived by determining the $\mathcal{G}_i$s of the involved groups G_i using Solid Edge. It can be verified that all of its m_i are lying in a virtual plane E_V symmetrically to the front and back loop of the robot. For most parallel robots with type III redundancies such a plane can be defined, especially for the planar ones.

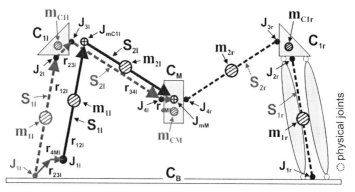

Figure 4.8: Simplified model with S_i, point masses and vector loops

Step 3 - dealing with coupling elements: The coupling elements C_{1l}, C_M and C_{1r} are substituted by point masses whose locations are determined with help of Solid Edge. Due to the symmetric construction of the SpiderMill m_{C1l} and m_{C1r} also lie in E_V. It has further been examined that effects of the unsymmetrically mounted end-effector mass m_{CM} are negligible. Therefore the end-effector is also assumed to be in E_V.

Step 3 (a) - joints of the simplified model: The axes of all rotational joints of the physical robot are perpendicular to E_V. Obviously, the movements of all S_is and all coupling elements C_is are limited within that plane, too. Since the real structure only contains elementary rotational joints ($f_i = 1$), all joints will be substituted by rotational joints with a DOF of one at the ends of the S_is.

Step 3 (b) - point mass(es) for the coupling elements: Analyzing the locations of the point masses m_{C1l}, m_{CM} and m_{C1r} with respect to the represented geometry of the rigid bodies (see Fig. 4.8) it is obvious that compared to the overall dimensions of the robot the distances δ_is, $(i = 1, \ldots, 3)$, especially δ_3, are very small (cp. Fig. 4.9 and 4.10) and the locations of the point masses are close to the vector loop which represents the kinematics of the simplified structure (see Fig. 4.8 - dashed line). Therefore, these distances δ_i are set to zero and the point masses are shifted to J_{mi}. Based on kinematic equivalence the geometrical parts representing the coupling elements (see Fig. 4.8 - solid line) and their belonging joints are further replaced by single joints J_{mi} (Fig. 4.9 and Fig. 4.10) - see also step 4 below.

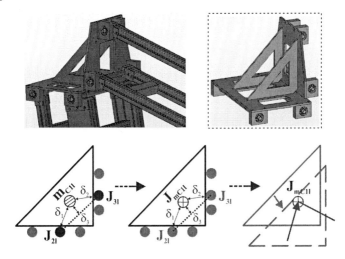

Figure 4.9: Substitution of the upper coupling elements C_{1k} (for the left side)

For safety reasons, the SpiderMill is equipped with an additional parking clamp allowing a safe mounting of the robot in its currentless resting state. In the context of our modeling strategy its mass can be interpreted as a high payload. This additional mass, assumed to be lying in E_V, is integrated in the calculation of m_{CM}. Although the distance δ_3 from the location of m_{CM} to the (dashed) vector loop increases thereby, it is further on set to zero for simplification. The practical results presented in Sec. 6.5.3 and Sec. 6.5.4 will show the correctness of this assumption.

Step 4 - statically determinate model - retaining of kinematics: The structure in Fig. 4.8 is statically indeterminate. For resting C_{1l} and C_{1r} the moving platform C_M e.g. can still be moved. This is not possible for the real physical

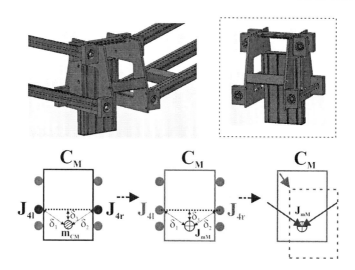

Figure 4.10: Substitution of the moving platform C_M

demonstrator. To arrive at a kinematically correct model, four vectors for each side of the manipulator are introduced (e.g. $\mathbf{r}_{23l}$, $\mathbf{r}_{4Ml}$, $\mathbf{r}_{12l}$, and $\mathbf{r}_{34l}$ for the left side). As C_{1l}, C_M and C_{1r} do not rotate relatively to the base coordinate system, $\mathbf{e}_{x,B}$, $\mathbf{e}_{y,B}$ in Fig. 4.11 – because of the parallel crank mechanisms of the robot, which constitute functionally type II redundancies – the dashed and the solid vector loop in Fig. 4.8 are kinematically equivalent. Therefore the three coupling elements can be shifted along the vector loop toward the base, resulting finally in a M-structure (Fig. 4.11). Their dynamic behavior can be represented by point masses as they cannot rotate relatively to the base coordinate system.

Step 5 - modeling of actuation: So far, the drive chains of the robot have not been considered. Now the ball screw systems actuating the axes (no direct driven joints) are added. The points of contact A_k (Fig. 4.11) are defined on the longitudinal axes of the substitution parts S_{1l} and S_{1r}. At the real demonstrator, cp. Fig. 4.7, the spindle bars are not mounted on the axes of symmetry of the links. Maintaining the distance d_A from the joint to the contact point there is a small offset. It has been verified that the resulting changes of the spindle orientations and therefore the effective directions of the spindle forces (the forces in the direction of the ball screw systems) can be neglected in the workspace. Because of the actuation concept all joints of SpiderMill are actually passive.

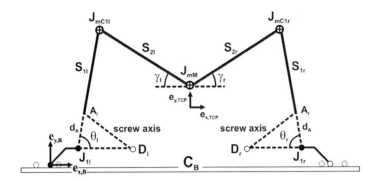

Figure 4.11: Simplified structure of the SpiderMill including actuation

4.3 Conclusions and future works

In this chapter a strategy for simplifying the structure of parallel robots with re-dundant links and passive joints, before deriving its dynamic equations has been presented. The approach is mainly based on the concept of dynamic equivalent masses and has been exemplarily applied to the planar, full-parallel robot Spi-derMill. Deriving its dynamic equations – based on several types of simplified structures – while applying different standard modeling approaches, the suitabil-ity and correctness of the proposed strategy will get obvious in Chap. 5.

5 Dynamic modeling

The derivation of a dynamic model of a robot is important in several ways. From control point of view it can be used as a simulation model in order to develop and test control strategies (including feedforward control) or it can also be part of the controller itself in case of model-based control concepts. Furthermore it is required for the planning of optimal trajectories. Beyond these aspects, the knowledge of the generalized joint forces can be used in the design phase in order to specify the link dimensions, the bearings and joint actuators.

Two types of dynamical problems can be distinguished. The *direct dynamic problem* (DDP) is to determine the response of the robot corresponding to the applied generalized forces. Which means that for a given vector of joint forces/torques the task is to compute the resulting motion of the robot as a function of time. In case of the *inverse dynamic problem* (IDP) the generalized actuator forces required to generate a desired trajectory of the end-effector have to be found. An efficient and accurate inverse dynamic model of a robot is essential for (real-time) feedforward control and model-based control strategies.

For the derivation of dynamics two different classes of strategies can be distinguished: variational or energy methods and geometry- or vector based (Newton-Euler). In the first case the extension of the principle of virtual work to dynamical problems so called D'Alembert's principle [Moo04] forms the base for common known concepts like Lagrange's equations [Tsa99, Cra05, SHV06], where all of the generalized coordinates are linearly independent. The Lagrangian approach eliminates the forces of constraint at the outset [Tsa99]. They have to be restored later if they are needed for design purpose. Both D'Alembert's method and Lagrange's equations can be used to formulate equations of motion considering geometric constraints (so called holonomic constraints). To also take velocity constraints or nonholonomic constraints into account principle of virtual power can be used directly. This strategy, also called Jourdain's principle [Jou09], can be considered as an extension of D'Alemberts principle or as a separate one. Also several variations of the method, such as Kane's equations [Kan61, KL85] exist. Furthermore Hamilton's principle, also a variational principle of dynamics, can be used to derive the dynamic equations of motion of multi-body systems [Sha05]. Using Newton-Euler method every rigid body of a mechanical system is treated separately by cutting free. The dynamic equations are derived solving the balances of forces and torques for all bodies. In two dimensional space three equations (2 translational, 1 rotational) and in three dimensional six equations (3 translational, 3 rotational) for each body have to be solved. In contrast to other holistic methods all coupling forces and torques become calculated (and are available) performing so called forward-backward recursion [SHV06, Cra05, Tsa99]. The recursive formulation of the algorithm is very effective for serial robots.

Since all of the here applied strategies for dynamic modeling of the SpiderMill are well known in literature, the final results are only sketched. The main purpose of this chapter is to show, that the strategy introduced in Chap. 4, can be applied in combination with all standard modeling approaches leading to good results. Nevertheless their theoretical background is shortly summarized in App. B.

This chapter is organized as follows: In Sec. 5.1 mainly analytical inverse dynamic models for the simplified structure of the SpiderMill (cp. Sec. 4.2) are derived applying several standard modeling approaches. Models for individual structural parts of the robot are given in Sec. 5.2. The general form of the equations of motion is discussed in Sec. 5.3 and the complexities of the derived analytical models of the SpiderMill are analyzed in Sec. 5.4. Different descriptions for the inverse and direct dynamic models of the SpiderMill are given in Sec. 5.5. Possibilities for the simulation of mechatronic systems are discussed in Sec. 5.6. Conclusions are drawn in Sec. 5.7.

5.1 Dynamic modeling of the SpiderMill

5.1.1 General considerations and simplified models

For deriving the equations of motion for the SpiderMill two kinds of generalized forces can be used as joint torques, the moments of the pseudo-active joints or the forces in spindle direction $\tau = [f_{sl} , f_{sr}]^T$. Here the latter are used. Furthermore it is assumed that no external torques τ_i^{ext} exist and in a first step the influence of friction is neglected while deriving the inverse dynamic models of the robot.

In Sec. 4.2 a simplified model of the SpiderMill – applying all steps of the proposed simplification strategy – has been introduced. For this especially the δ_is have been set to zero as a result of simplification step 3. The obtained structure, shown in Fig. 4.11, will be used in Sec. 5.1.3 for deriving the equations of motion with the Newton-Euler approach.

Beside this simplified structure another one only applying steps 1, 2 and 5 in the explained manner and replacing all coupling elements by point masses located in their centers of masses (step 3) – but not shifting the coupling elements down in order to obtain a M-structure of the robot (step 4 is not performed) – is additionally introduced. This second model, shown in Fig. 5.1, is always used unless otherwise told. As a consequence the δ_is become part of the set up equations, but not necessarily of the final solutions. The results below will show, that the statical determinateness of the model has not necessarily be realized by the simplified structure (black in Fig. 5.1) itself (e.g. by shifting of coupling elements). Rather it can also be realized in an implicit manner based on the kinematic relationships of the physical structure (gray). This means the locations of the point masses (obtained by applying the simplification strategy – without step 4) are taken into account, while deriving the equations of motion. But their correct motion characteristic is enforced by the parametrization of the equations in dependency of the end-effector movement (which is governed by the kinematics of the physical structure).

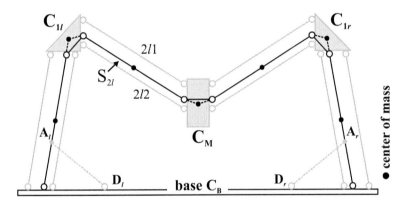

Figure 5.1: Simplified base model of the SpiderMill

In order to verify the analytically derived rigid body models a frictionless Sim-Mechanics model of the SpiderMill will be used. For its implementation the same rigid body parameters – taken from CAD construction data – as in case of the MSC.ADAMS model and for the parameterization of the analytical equations (cp. Table A.2) are used. Since the physical structure in Fig. 5.1 (dynamically underdetermined) cannot be solved using SimMechanics the structure shown in Fig. 2.6 (and Fig. 5.4) has been implemented. As verification trajectories the circle trajectories planned in Sec. 7.3 for a jerk limit[1] of 1.6 Nm s^{-1} will be used as inputs

---

[1]In general the unit of the jerk is $m\ s^{-3}$. But in this thesis the term jerk is used with respect to the time derivative of the motor torques.

for the spindle motions. The model comparison will be performed based on the spindle forces. The results for all approaches are summarized in compact form in Tab. 5.1.

In the following the most typical concepts for the dynamic modeling of (parallel) robots are applied in order to derive the equations of motion of the SpiderMill. A detailed discussion of the approaches is given in App. B.

5.1.2 Lagrangian approaches

The first step of applying the Lagrangian approaches to a mechanical system is to derive its kinetic and potential energy. In case of the SpiderMill the kinetic energy is defined by

$$
\begin{aligned}
\mathcal{T} =& \frac{1}{2}m_{1l}\left(\dot{x}_{m1l}^2 + \dot{y}_{m1l}^2\right) + \frac{1}{2}m_{2l}\left(\dot{x}_{m2l}^2 + \dot{y}_{m2l}^2\right) + \frac{1}{2}m_{3l}\left(\dot{x}_{m3l}^2 + \dot{y}_{m3l}^2\right) \\
&+ \frac{1}{2}m_{CM}\left(\dot{x}_{mCM}^2 + \dot{y}_{mCM}^2\right) + \frac{1}{2}m_{3r}\left(\dot{x}_{m3r}^2 + \dot{y}_{m3r}^2\right) + \frac{1}{2}m_{2r}\left(\dot{x}_{m2r}^2 + \dot{y}_{m2r}^2\right) \\
&+ \frac{1}{2}m_{1r}\left(\dot{x}_{m1r}^2 + \dot{y}_{m1r}^2\right) + \frac{1}{2}J_1\dot{\theta}_l^2 + \frac{1}{2}J_3\dot{\gamma}_l^2 + \frac{1}{2}J_3\dot{\gamma}_r^2 + \frac{1}{2}J_1\dot{\theta}_r^2
\end{aligned}
\tag{5.1}
$$

where $(x_{mik},\, y_{mik})$ are the coordinates of the center of mass $\mathcal{G}_i$ of mass i of side k (or the moving platform CM) defined with respect to the base coordinate system $(E)_B$. $(\dot{x}_{mik},\, \dot{y}_{mik})$ are the corresponding velocities. The potential energy is given by

$$
\begin{aligned}
\mathcal{U} =& m_{1l}\,g\,y_{m1l} + m_{2l}\,g\,y_{m2l} + m_{3l}\,g\,y_{m3l} + m_{CM}\,g\,y_{mCM} + m_{3r}\,g\,y_{m3r} \\
&+ m_{2r}\,g\,y_{m2r} + m_{1r}\,g\,y_{m1r}.
\end{aligned}
\tag{5.2}
$$

Lagrangian equations of the second type (Lagrange II)

For the SpiderMill in case of the Lagrangian approach of the second type (cp. App. B.2.1 for more details) the vector of the generalized joint variables $\mathbf{q}$ contains only the pseudo-active coordinates

$$
\mathbf{q} = \begin{bmatrix} \theta_l, & \theta_r \end{bmatrix}^T.
\tag{5.3}
$$

The passive coordinates – occurring in the equations of the first type – become expressed by the pseudo-active ones according to Eq. (3.13). Based on it the Lagrangian equations for the SpiderMill can be set up as

$$
\boldsymbol{\tau}_l = \frac{d}{dt}\left(\frac{\partial \mathcal{L}}{\partial \dot{\theta}_l}\right) - \frac{\partial \mathcal{L}}{\partial \theta_l} \qquad \text{and} \qquad \boldsymbol{\tau}_r = \frac{d}{dt}\left(\frac{\partial \mathcal{L}}{\partial \dot{\theta}_r}\right) - \frac{\partial \mathcal{L}}{\partial \theta_r}.
\tag{5.4}
$$

The advantage of this approach compared to that of the first type is that the DDP can be solved in Matlab. In case of the first type problems regarding the initialization of the numerical solvers occur[2].

The results of the comparison of the analytically derived inverse dynamics with the SimMechanics model of the SpiderMill are summarized in Tab. 5.1. Whereas the small occurring errors justify the assumption of correct performed modeling steps and consequently that of the modeling approach introduced in Chap. 4. This does also hold for all the other, in the following introduced approaches.

Lagrangian equations of the first type (Lagrange I)

In case of the Lagrangian approach of the first type (cp. App. B.2.2) $\mathbf{q}$ contains beside the pseudo-active also the passive coordinates of the SpiderMill.

$$\mathbf{q} = \begin{bmatrix} \theta_l, & \theta_r, & \gamma_l, & \gamma_r \end{bmatrix}^T \tag{5.5}$$

Because of the two passive coordinates two Lagrangian multipliers λ_j and two (kinematic) constraint functions Γ_j are required. The constraint functions are derived based on the closed vector-loop of the robot. With the chains of the left and right side, cp. Eq. (3.3) and (3.5) the constraint functions can be defined to

$$\Gamma_1 = \Gamma_{1l}^* - \Gamma_{1r}^* = 0 \quad \text{and} \quad \Gamma_2 = \Gamma_{2l}^* - \Gamma_{2r}^* = 0. \tag{5.6}$$

Based on the definitions above the Lagrangian equations of the first type are

$$0 = \frac{d}{dt}\left(\frac{\partial \mathcal{L}}{\partial \dot{\gamma}_l}\right) - \frac{\partial \mathcal{L}}{\partial \gamma_l} - \sum_{j=1}^{2} \lambda_j \frac{\partial \Gamma_j}{\partial \gamma_l} \tag{5.7a}$$

$$0 = \frac{d}{dt}\left(\frac{\partial \mathcal{L}}{\partial \dot{\gamma}_r}\right) - \frac{\partial \mathcal{L}}{\partial \gamma_r} - \sum_{j=1}^{2} \lambda_j \frac{\partial \Gamma_j}{\partial \gamma_r} \tag{5.7b}$$

$$\boldsymbol{\tau}_l = \frac{d}{dt}\left(\frac{\partial \mathcal{L}}{\partial \dot{\theta}_l}\right) - \frac{\partial \mathcal{L}}{\partial \theta_l} - \sum_{j=1}^{2} \lambda_j \frac{\partial \Gamma_j}{\partial \theta_l} \tag{5.7c}$$

$$\boldsymbol{\tau}_r = \frac{d}{dt}\left(\frac{\partial \mathcal{L}}{\partial \dot{\theta}_r}\right) - \frac{\partial \mathcal{L}}{\partial \theta_r} - \sum_{j=1}^{2} \lambda_j \frac{\partial \Gamma_j}{\partial \theta_r}. \tag{5.7d}$$

The left sides of Eq. (5.7a) and (5.7b) are set to zero because it is assumed that no external torques $\boldsymbol{\tau}_i^{ext} = 0$ act.

---

[2]The mass matrices in case of both approaches can become (configuration dependently) singular. But especially in the initial state the mass matrix of the first type is singular and therefore allows the numerical solver not to start.

5.1.3 Newton-Euler approach

Newton-Euler approach for the simplified structure

In this section the Newton-Euler approach (cp. App. B.3) is applied in combination with the simplification strategy proposed in Sec. 4.2. The obtained model defines the basis for the following chapters and is used as dynamic model of the SpiderMill. For that reason it is, in contrast to the others discusssed in more detail and in Sec. 6.1.2 extended by the friction torques of the passive joints and the friction losses in the drive systems.

Using the proposed modeling strategy the 17 bodies (cross braces not counted) of the physical robot are replaced by four S_is, reducing largely the number of system equations from 102 to 12. The four S_is, cp. Fig. 4.8 and 4.11, are cut free and the generalized velocities and accelerations are calculated under considerations of Sec. 4.2. In the following the procedure is exemplarily shown for S_{1l} in Fig. 5.2. Its balances of forces and moments are given in Eq. (5.8a) and (5.8b)

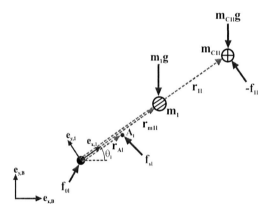

Figure 5.2: Forces and moments on substitution part S_{1l} (and S_{2l})

$$^B\mathbf{f}_{0l} + {}^B\mathbf{f}_{sl} - {}^B\mathbf{f}_{1l} + (0 \quad -m_1 g \quad 0)^T + (0 \quad -m_{C1l}g \quad 0)^T = m_1\,{}^B\dot{\mathbf{v}}_{m1l} + m_{C1l}\,{}^B\dot{\mathbf{v}}_{mC1l}$$
$$(5.8a)$$

$$^B\mathbf{r}_{Al} \times {}^B\mathbf{f}_{sl} + {}^B\mathbf{r}_{m1l} \times (0 \quad -m_1 g \quad 0)^T + {}^B\mathbf{r}_{1l} \times (0 \quad -m_{C1l}g \quad 0)^T + {}^B\mathbf{r}_{1l} \times (-{}^B\mathbf{f}_{1l})$$
$$= m_1\,{}^B\mathbf{r}_{m1l} \times {}^B\dot{\mathbf{v}}_{m1l} + m_{C1l}\,{}^B\mathbf{r}_{1l} \times {}^B\dot{\mathbf{v}}_{mC1l} + {}^B\mathbf{I}_1^{c1}\,{}^B\dot{\boldsymbol{\omega}}_{1l} + {}^B\boldsymbol{\omega}_{1l} \times ({}^B\mathbf{I}_{1l}^{m1l}\,{}^B\boldsymbol{\omega}_{1l}).$$
$$(5.8b)$$

Due to the planar M-structure of the robot only three scalar equations are contained in the balance equations. While cutting free the substitution elements S_{1l} and S_{2l} the whole mass of m_{C1l} is added to S_{1l} (located in J_{mC1l}). This step is motivated by practical considerations. By moving C_{1l}, the element S_{1l} (or the respective groups of the real demonstrator) is rotating with respect to the substitution joint J_{1l}. The distances between m_{C1l} and the base joints of physical demonstrator are nearly equal to the length d_{1l} of S_{1l}, having the function of a lever arm for m_{C1l}. To accelerate m_{C1l} substitution part S_{1l} has to be actuated. The general balance equations are solved for

$$\boldsymbol{\tau} = \begin{bmatrix} \mathbf{f}_{sl}, & \mathbf{f}_{sr} \end{bmatrix}^T = \boldsymbol{\tau}_{rb}. \tag{5.9}$$

The comparison of the analytical and simulated inverse dynamics is exemplarily shown in Fig. 5.3. The relatively small deviations from the SimMechanics model

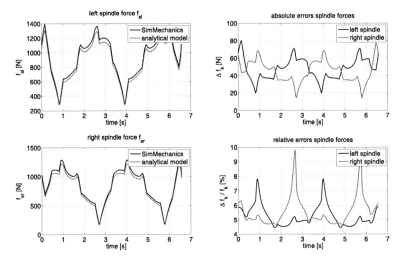

Figure 5.3: Spindle forces and errors in case of reduced Newton-Euler model

demonstrate clearly the correctness of the proposed modeling strategy in case of SpiderMill. The errors of the simplified model mainly result from the chosen locations for the point masses representing the upper coupling elements C_{1l}, C_{1r} and the moving platform C_M, cp. Sec. 4.2 for details. In Sec. 6.5 the model becomes further improved by identifying its parameters instead of using values calculated from CAD data.

Newton-Euler approach for the 'full' structure

In order to further analyze the assumptions of the simplified modeling strategy
in Sec. 4.2 also a dynamic model of the only planarized robot is necessary. A
complete 3-dimensional model will not be derived, since also both implemented
MBS models of the SpiderMill are planar. The 3-dimensionality shown in Fig. 4.7
is only a visualization feature of the MSC.ADAMS model.

A modeling of the SpiderMill based on the structure shown in Fig. 5.1, with four
parallel crank mechanisms would lead to a dynamically under-determined system of
equations. In order to avoid the problems finding a solution of it (which can exist,
but which does not necessarily have to exist), the model is reduced to the structure
shown in Fig. 5.4. This structure will be obtained by applying step 2 of the
simplification strategy on the links $2l1$ and $2l2$ ($\mapsto S_{2l}$) of the robot. In principle,
also other redundancies could have been eliminated in order to reduce the resulting
system of equations to a dynamically determined and therefore uniquely solveable
one. Here this redundancy has been chosen in order to match the analytical with
the SimMechanics model, cp. Fig. 2.6, which has been reduced in the same way
in order to allow Matlab/Simulink to solve its equations of motion.

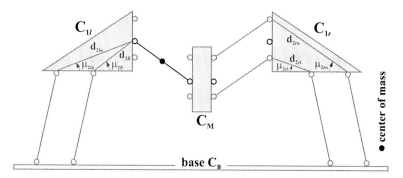

Figure 5.4: Solvable model of the robot

The procedure for solving the IDP is the same as in case of the simplified structure. The comparison of the analytically derived and simulated inverse dynamics
is exemplarily shown in Fig. 5.5.

Analyzing the obtained equations of motion in more detail the correctness of
the proposed simplified modeling strategy (cp. Sec. 4) is further confirmed[3]. In

---

[3]Since the same aspects get even better obvious from the analysis in Sec. 5.2.1 and 5.2.2 ana-

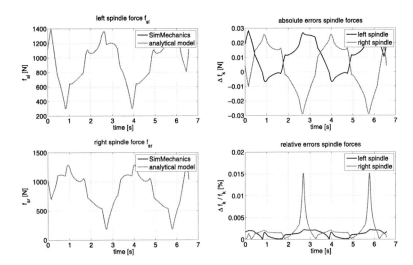

Figure 5.5: Spindle forces and errors in case of Newton-Euler approach

the equations the lumped mass approach can be applied to the masses and inertia terms of the single links of the robot, leading to the simplification proposed in Sec. 4.2. This means the inner and outer links defining the parallel crank mechanism can be combined in the proposed way. A further interesting aspect analyzing the final equations is the fact, that the additionally introduced kinematic parameters (the μ_{2i}s in Fig. 5.4 and also the d_{2i}s) do not appear. This point will be further discussed in Sec. 5.2. Due to these results the errors in case of the simplified structure (see above) can only be caused by the rearrangement of the elements (step 4) leading to the structure in Fig. 4.11.

5.1.4 Variational principles - Virtual power

Since virtual work and virtual power are only two different interpretations of the same basic principle only the results for the latter one will be presented in this section. Applying the straight forward approach introduced in [Gro03,AGH05,Abd07] (cp. also App. B.4.2) for parallel robots the final equations for the SpiderMill with

---

lyzing subsystems of the full structure, this results are not presented here.

its $N_k = 7$ bodies (simplified structure) are given by

$$
\boldsymbol{\tau}_{rb} = \sum_{i=1}^{3} \begin{bmatrix} \mathbf{J}_{T_{i,l}}^T & \mathbf{J}_{R_{i,l}}^T \end{bmatrix} \boldsymbol{\Omega}_{i,l} \begin{bmatrix} \mathbf{I}_{i,l}^\diamond \\ \mathbf{s}_{i,l} \\ m_{i,l} \end{bmatrix} + \begin{bmatrix} \mathbf{J}_{T_{CM}}^T & \mathbf{J}_{R_{CM}}^T \end{bmatrix} \boldsymbol{\Omega}_{CM} \begin{bmatrix} \mathbf{I}_{CM}^\diamond \\ \mathbf{s}_{CM} \\ m_{CM} \end{bmatrix}
$$

$$
+ \sum_{i=1}^{3} \begin{bmatrix} \mathbf{J}_{T_{i,r}}^T & \mathbf{J}_{R_{i,r}}^T \end{bmatrix} \boldsymbol{\Omega}_{i,r} \begin{bmatrix} \mathbf{I}_{i,r}^\diamond \\ \mathbf{s}_{i,r} \\ m_{i,r} \end{bmatrix} \tag{5.10}
$$

The results of the comparison with the simulated inverse dynamics are summarized in Tab. 5.1.

5.1.5 Comparison of the approaches

In order to compare the different modeling approaches the absolute values of the mean absolute (MAE) and relative (MRE) errors are summarized in Table 5.1. It seems that the only reason, why the results of the reduced Newton-Euler model (as only model resting upon the structure in Fig. 4.11) are not as perfect as that of the other models is the shifting of the coupling elements performed in step 4 of the simplification strategy. But also this model is well suited for the steps in the following chapters on the way to a model-based trajectory tracking control. Generally, the small deviations between the calculated and simulated forces clearly demonstrate the correctness of the modeling strategy proposed in Chap. 4 in case of the SpiderMill.

approach	$|\text{MAE}_l|$ [N]	$|\text{MAE}_r|$ [N]	$|\text{MAE}_{lr}|$ [N]	$|\text{MRE}_l|$ [%]	$|\text{MRE}_r|$ [%]	$|\text{MRE}_{lr}|$ [%]
Lagrange I	0.0488	0.0404	0.0446	0.0054	0.0044	0.0049
Lagrange II	6.3802	6.5515	6.4658	0.8115	1.0205	0.9160
Newton-Euler (sim.)	47.4567	44.8266	46.1417	5.2634	5.5887	5.4260
Newton-Euler (full)	0.0126	0.0139	0.0133	0.0012	0.0022	0.0017
Virtual power	0.0488	0.0403	0.0445	0.0054	0.0044	0.0049

Table 5.1: Comparison of the absolute values of the absolute and relative errors

5.2 Reduced models of the SpiderMill

In Sec. 5.1 models for the complete SpiderMill have been derived. These models are necessary in order to develop model-based control strategies in Chap. 10, since they describe the dynamic behavior of the entire robot.

In this section three additional models are introduced. Since they only describe parts of the SpiderMill they are named 'reduced models' in the following. The first two of them (cp. Sec. 5.2.1 and 5.2.2) are used to further evaluated the model simplification strategy of Chap. 4 by analyzing structural parts of the SpiderMill. Thereto one analytical model is derived directly for the structural part, without applying the steps of the simplification strategy in advance (called 'full' model - FM). Additionally a second model is derived after applying the simplification steps (called 'simplified' model - SM). Both are compared to a SimMechanics model (SMM) of the structural part and also with each other. The results of the comparison of the dynamic equations of the full and the simplified analytical model allow a further verification of the simplifications steps (cp. Chap. 4) performed in order to derive the models of the complete SpiderMill in Sec. 5.1.

The third discussed model (cp. Sec. 5.2.3) forms the basis for one of the non-model-based control strategies in Sec. 10.5. First it is compared to an equivalent SimMechanics model and in a second step to the SimMechanics model of the whole robot.

For all three models the absolute values of the MAE and MRE are summarized in Table 5.2 in a compact form. As reference trajectory the same as in Sec. 5.1 is chosen.

5.2.1 Dynamics of the left parallel crank mechanism

The equations of motion of the left parallel crank mechanism and its simplified model, cp. Fig. 5.6 (actuation not sketched), are analyzed in the following in order

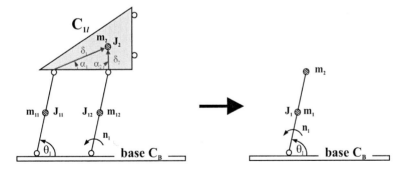

Figure 5.6: Left parallel crank mechanism and simplified model

to allow a further discussion of the assumptions of the simplified modeling strategy
(cp. Sec. 4.2) and the particularities observed in Sec. 5.1.3. The final dynamic
equation of the parallel crank mechanism is

$$
\begin{aligned}
\mathbf{n}_l = {} & \left(J_{11} + J_{12} \right) \ddot{\theta}_l + d_{m1}^2 \left(m_{11} + m_{12} \right) \ddot{\theta}_l + d_1^2 \, m_2 \, \ddot{\theta}_l \\
& + g \, d_{m1} \left(m_{11} + m_{12} \right) \cos \theta_l + d_1 \, m_2 \, g \, \cos \theta_l
\end{aligned}
\tag{5.11}
$$

and that of its simplified model

$$
\mathbf{n}_l = J_1 \ddot{\theta}_l + d_{m1}^2 m_1 \ddot{\theta}_l + d_1^2 \, m_2 \, \ddot{\theta}_l + g \, d_{m1} m_1 \cos \theta_l + d_1 \, m_2 \, g \, \cos \theta_l.
\tag{5.12}
$$

The equations Eq. (5.11) and Eq. (5.12) are equal for

$$
m_{11} + m_{12} = m_1 \qquad \text{and} \qquad J_{11} + J_{12} = J_1.
\tag{5.13}
$$

This corresponds exactly to the result, when the simplified modeling strategy (cp.
Sec. 4.1.4 and 4.2) is applied. In Fig. 5.7 the spindle forces of both analytical
models are compared to the forces determined from the SimMechanics model of
the left parallel crank mechanism.

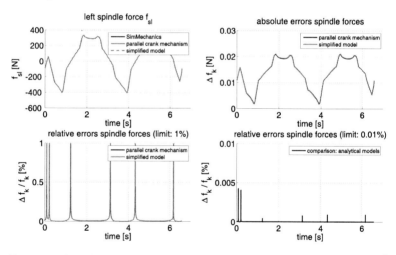

Figure 5.7: Spindle forces and errors in case of the left parallel crank mechanism[4]

---

[4]In order to show more of the relevant parts of the plots of the relative and absolute errors and
not only the size of a few peaks (only occurring due to numerical effects) some of the plots
are cut off at the denoted limit values in the following.

The results suggest the correctness of the analytical models. The also performed direct comparison of both analytical models show, that the peaks of the relative error are a numerical phenomenon, occurring at the zero crossings of the moment. A further interesting result is the fact, that the leaver arms δ_i and its corresponding angles α_i do not occur in the equation of the full model (5.11). In order to further investigate their influence the dynamic equation of the left limb is derived in Sec. 5.2.2.

5.2.2 Dynamics of the left limb

In a further step the dynamic equations of the left limb of the SpiderMill and its simplified model in Fig. 5.8 are analyzed. This step is performed in order to further

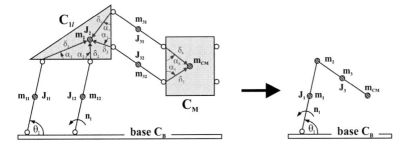

Figure 5.8: Left limb and simplified model

investigate if there is an influence of the δ_i and the corresponding angles α_i on the system dynamics. The dynamic equation of the simplified model is

$$
\begin{aligned}
\mathbf{n}_l =\ & g\left(d_{m1}m_1 + d_1\left(m_2 + m_3 + m_{CM}\right)\right)\cos\theta_l + \left(J_1 + d_{m1}^2 m_1\right)\ddot{\theta}_l \\
& + d_1\left(\left(d_{m3}m_3 + d_3 m_{CM}\right)\left(\sin\left(\gamma_l + \theta_l\right)\dot{\gamma}_l^{\,2} - \cos\left(\gamma_l + \theta_l\right)\ddot{\gamma}_l\right)\right. \\
& \left. + d_1\left(m_2 + m_3 + m_{CM}\right)\ddot{\theta}_l\right).
\end{aligned}
\tag{5.14}
$$

Like in case of the parallel crank mechanism (cp. Sec. 5.2.1) also for the whole left limb, the δ_is and α_is do not occur in the analytical equation (5.14) of the full model. The difference between the simplified and full model is given by

$$
\begin{aligned}
\Delta n_l =\ & \frac{d_1}{d_3}\cos\theta_l\sec\gamma_l\left(d_1\left(d_{m3}m_3 + d_3 m_{CM}\right)\sin\left(\gamma_l + \theta_l\right)\dot{\theta}_l^{\,2} + \left(J_3 + d_{m3}^2 m_3\right.\right. \\
& \left. + d_3^2 m_{CM}\right)\ddot{\gamma}_l - \left(d_{m3}m_3 + d_3 m_{CM}\right)\left(g\cos\gamma_l + d1\cos\left(\gamma_l + \theta_l\right)\ddot{\theta}_l\right)\right)
\end{aligned}
\tag{5.15}
$$

From the comparison of the spindle forces of both analytical models shown in Fig. 5.9, it can be concluded, that the influence of Eq. (5.15) is – at least in case of the reference trajectory – only marginal. The absolute values of the MAE and

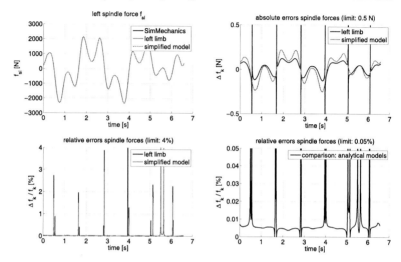

Figure 5.9: Spindle forces and errors in case of the left limb

MRE are given in Table 5.2.

Very interesting is the fact that neither the μ_{2i}s and d_{2i}s in case of the full dynamic model of the SpiderMill (cp. Sec. 5.1.3) nor the α_is and δ_is in case of the full model of the left parallel crank mechanism (cp. Sec. 5.2.1) and limb occur in the respective analytical equations. Due to these results it seems to be justified to neglect those lever arms (like d_{2i}s and δ_is) while deriving the equations of motion in order to simplify the analytical modeling of parallel robots with redundancies.

5.2.3 Single link with end-mass

In this section a reduced model of the SpiderMill in form of a single axis model with end-mass is derived. The obtained model is identical for both limbs of the robot, whereas the mass of the moving platform m_{CM} is cut into halves. The dynamic equation for the model in Fig. 5.10 is given by

$$\mathbf{n}_k = \left(I_{m12} + d_{m1}^2\, m_1 + d_1^2\, m^*\right) \ddot{\theta}_k + g\,(d_{m1}\, m_1 + d_1\, m^*)\cos\theta_k \qquad (5.16)$$

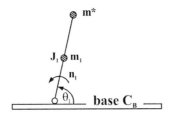

Figure 5.10: Single axis with end-mass m^*

whereas the point-mass m^* is defined by

$$m^* = m_2 + m_3 + \frac{1}{2}m_{CM}. \qquad (5.17)$$

The results of the comparison of the analytical model and a MBS model of the structure in Fig. 5.10 are given in Fig. 5.11. The also shown comparison with the

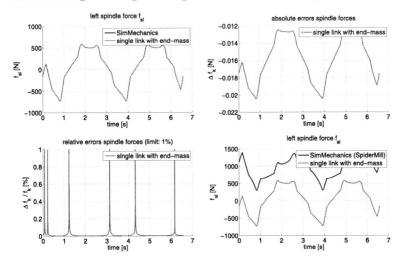

Figure 5.11: Spindle forces and errors in case of a single link with end-mass

full model of the SpiderMill justifies the derivation of the analytical full dynamic models in Sec. 5.1.

type of model	type of error	SMM $\leftrightarrow$ FM	SMM $\leftrightarrow$ SM	FM $\leftrightarrow$ SM
left parallel	|MAE| [N]	0.0134	0.0135	$1.2289 \cdot 10^{-5}$
crank	|MRE| [%]	0.0338	0.0338	$3.5223 \cdot 10^{-5}$
left limb	|MAE| [N]	0.0313	0.0177	0.0136
	|MRE| [%]	0.0302	0.0260	0.0120
link with	|MAE| [N]	0.0158	–	–
end-mass	|MRE| [%]	0.0279	–	–

Table 5.2: Comparison of the relative and absolute errors of the reduced models

Final result: In Table 5.2 all derived reduced analytical models are compared with their MBS simulation in SimMechanics. The results strongly justify the proposed modeling strategy and its simplification steps. Furthermore it is shown, that the single link with end-mass model of the SpiderMill can in no case keep up with the full models derived in Sec. 5.1.

5.3 General form of the dynamic equations

In order to give the equations of motion in a compact form, hiding some details, but allowing an easier and also physical interpretation the following form is commonly chosen

$$\mathbf{M}\left(\mathbf{q}\right)\ddot{\mathbf{q}} + \mathbf{C}\left(\mathbf{q},\dot{\mathbf{q}}\right) + \mathbf{G}\left(\mathbf{q}\right) = \boldsymbol{\tau}. \tag{5.18}$$

Eq. (5.18) is known as the general form of dynamic equations (in the joint space) or Lagrangian form. The mass matrix $\mathbf{M}\left(\mathbf{q}\right)_{(n \times n)}$ accounts the inertia forces of the manipulator, it is always a symmetric and positive definite matrix and therefore invertible [Tsa99][5]. Its diagonal elements represent the effective inertias at the respective joints [Dor97], while the off-diagonal terms represents the acceleration coupling effects between the joints (coupling inertias). The velocity coupling vector $\mathbf{C}\left(\mathbf{q},\dot{\mathbf{q}}\right)_{(n \times 1)}$, represents the centrifugal and Coriolis terms. Elements depending on the squares of the velocities $\dot{q}_i^2$ are caused by the centrifugal forces, whereas terms depending on a velocity product $\dot{q}_i \cdot \dot{q}_j$ (for $i \neq j$) are caused by a Coriolis force. Also the influence of the friction forces $\boldsymbol{\tau}_f$ can be included in the closed form model

$$\mathbf{M}\left(\mathbf{q}\right)\ddot{\mathbf{q}} + \mathbf{C}\left(\mathbf{q},\dot{\mathbf{q}}\right) + \mathbf{G}\left(\mathbf{q}\right) + \boldsymbol{\tau}_f\left(\dot{\mathbf{q}}\right) = \boldsymbol{\tau}. \tag{5.19}$$

---

[5]This holds in most cases, but not always – cp. Sec. 5.1.2 and [ER94].

The choice of $\mathbf{C}\left(\mathbf{q},\dot{\mathbf{q}}\right)$ is not unique [SS05]. In order to get one with special properties the variant with the Christoffel symbols of the first type can be set-up, reducing the computational effort calculating $\mathbf{C}\left(\mathbf{q},\dot{\mathbf{q}}\right)$ [SHV06]. The vector of gravitational forces $\mathbf{G}\left(\mathbf{q}\right)_{(n\times1)}$ represents the gravitational effects. The element on the right side of the equation is the $(n\times1)$ vector of the generalized forces. It is discussed in Sec. B.5 in detail. Beside the standard form of Eq. (5.18) also other variants like the following exist. By rearranging the velocity dependent term $\mathbf{C}$, the dynamic equation can be written as [Cra05]

$$\mathbf{M}\left(\mathbf{q}\right)\ddot{\mathbf{q}} + \mathbf{B}\left(\mathbf{q}\right)\left[\dot{\mathbf{q}}\dot{\mathbf{q}}\right] + \mathbf{V}\left(\mathbf{q}\right)\left[\dot{\mathbf{q}}^2\right] + \mathbf{G}\left(\mathbf{q}\right) = \boldsymbol{\tau} \tag{5.20}$$

where $\mathbf{B}\left(\mathbf{q}\right)$ is the matrix of Coriolis coefficients $(n\times n(n-1)/2)$ and $\left[\dot{\mathbf{q}}\dot{\mathbf{q}}\right]$ the vector of joint velocity products $(n(n-1)/2\times1)$ given by

$$\left[\dot{\mathbf{q}}\dot{\mathbf{q}}\right] = \left[\dot{q}_1\dot{q}_2 \ \dot{q}_1\dot{q}_3 \ \ldots \ \dot{q}_{n-1}\dot{q}_n\right]^T \tag{5.21}$$

$\mathbf{V}\left(\mathbf{q}\right)$ is the matrix of centrifugal coefficients $(n\times n)$ and $\left[\dot{\mathbf{q}}^2\right]$ the vector of joint velocity square products $(n\times1)$ given by

$$\left[\dot{q}_1^2 \ \dot{q}_2^2 \ \ldots \ \dot{q}_n^2\right]^T. \tag{5.22}$$

Eq. (5.20) is called configuration-space equation [Cra05, RH78], because all matrices are functions of the robot posture or the joint coordinates, respectively.

The fact that the matrices of the different formulations of the equations of motion above are only depending on $\mathbf{q}$ and its first time derivative is important, when realizing control concepts for a robot. It allows to update the dynamic model depending on how fast the configuration of a manipulator is changing moving on a trajectory in order to reduce the computational load.

5.4 Complexity and interpretation of the full dynamic models

Even in case of serial robots the complexity of the dynamic models is a critical point. This does hold even more for parallel robots, especially when model-based real-time control is the goal. For serial robots a lot of research has been done [Hol80, Cra05] comparing different modeling approaches and their computational load. Thereby the reasons for the effectivity or non-effectivity of an approach are related to the structure of the obtained equations of motion and the way how they are solved. For parallel robots only a few comparisons of the equations of

motion derived with different approaches for the same manipulator can be found –
whereas most of it only consider the Stewart-Gough platform [ST98, Tsa99]. The
obtained dynamic models cannot be transformed into each other due to existing
coupling terms, resulting from the applied methods. Furthermore, they are not all
parameter-minimal but can normally be reduced to this form, applying analytical
and/or numerical methods [GK86, GK90, GH00, Gau90]. Also in case of parallel
robots, strategies for deriving dynamic modeling of a lower complexity have been
proposed by several authors [Lin90, PKB+05, LLL93, Sta97, KML09].

An important point is to distinguish closed form and iterative formulations of the
dynamic equations. The second ones are very interesting from the simulation view-
point, whereas the first ones allow a direct application of standard model-based
control algorithms. But as discussed in [Cra05] closed-form dynamic equations
derived for a particular manipulator can even be simulated more effective than
applying a general iterative scheme. The main drawback of closed-form equations
is the effort to derive them. Several special software tools are available assisting at
this step, but also a standard computer algebra system (CAS: e.g. Mathematica
or Maple) – like used here – can be applied. It is also possible to get a dynamic
model based on object oriented [HSvS05, H05, OEC96, HS04] techniques or a mix
between symbolic and numerical methods [GST+05]. Since in case of the Spider-
Mill the equations of motion are not derived to get a simulation model (DDP) the
approaches for solving them effectively are not discussed.

In Sec. 5.1 dynamic models have been derived for the SpiderMill based on several
approaches. In order to be able to compare their complexity the individual matrices of their state-space equations, cp. Eq. (5.18), have to be defined in an identical
manner. For that purpose all models become expressed in the following form

$$
\begin{bmatrix} M_{11} & M_{12} \\ M_{21} & M_{22} \end{bmatrix} \cdot \begin{bmatrix} \ddot{\theta}_l \\ \ddot{\theta}_r \end{bmatrix} + \begin{bmatrix} C_{11} & C_{12} & C_{13} \\ C_{21} & C_{22} & C_{23} \end{bmatrix} \cdot \begin{bmatrix} \dot{\theta}_l \cdot \dot{\theta}_l \\ \dot{\theta}_r \cdot \dot{\theta}_r \\ \dot{\theta}_l \cdot \dot{\theta}_r \end{bmatrix} + \begin{bmatrix} G_{11} \\ G_{21} \end{bmatrix} = \begin{bmatrix} f_l \\ f_r \end{bmatrix} \qquad (5.23)
$$

whereas all elements of the matrices $\mathbf{M}$, $\mathbf{C}$ and $\mathbf{G}$ are at most functions of the
passive joint angles γ_k.

In literature the complexity of a closed form model is in most cases evaluated by
the number of multiplications and additions and sometimes the powers of elements
(if not replaced by binary operations). But in realtiy, the number of trigonometric calculations is even more critical for the implementation of real-time control
strategies, strongly limiting the achievable sampling rates. For that reason also
their number is considered while evaluating the complexity of the derived models
of the SpiderMill. In order to do the comparison in a compact form Table 5.3 is
given, where t_{ex} stands for the execution time of the corresponding Matlab code.
From Table 5.3 it is obvious, why in case of model-based control concepts at most

approach	matrices	binary op.	a^b	trigonometric op.
Lagrange 1	M	321	65	26
	V	852	115	124
	G	245	115	28
$t_{ex} = 12.41$ s	MVG	1 418	295	178
Lagrange 2	M	103 469	15 833	10 958
	V	687 150	102 006	71 958
	G	16 202	2 486	1 723
$t_{ex} = 1\,637.36$ s	MVG	806 821	120 325	84 639
Newton-Euler	M	442	70	38
(simplified model)	V	4 320	280	358
	G	118	14	28
$t_{ex} = 13.12$ s	MVG	4 880	364	424
Newton-Euler	M	395	64	26
(full model)	V	4 071	320	538
	G	118	14	22
$t_{ex} = 21.57$ s	MCG	4 584	398	586
Virtual power	M	88 298	31 148	16 692
	V	131 618	46 302	25 050
	G	34 524	12 068	6 546
$t_{ex} = 85.68$ s	MVG	254 440	89 518	48 288

Table 5.3: Comparison of the numerical complexity of the approaches

the vector of the gravitational forces $\mathbf{G}\,(\mathbf{q})$ is calculated online.

Comparing the model complexities based on the values summarized in Table 5.3 it has to be kept in mind, that these values are not those for the most compact coding of the equations of motion. Carrying out for example the calculation of the time derivatives of the passive joint angles and also that of their partial derivatives only one time the complexity of some of the models can strongly be reduced. But the values in Table 5.3 have been obtained by replacing the above mentioned variables by algebraic functions of the pseudo-active and passive coordinates, followed by an application of the standard simplification routines of Mathematica.

All of the above introduced full dynamic models of the SpiderMill are appropriate for the steps in the next chapters. Without a doubt, the best analytical model, with respect to its complexity and execution time, is obtained by applying the Lagrangian approach of the first type. The apparently inconsistently choice of the reduced Newton-Euler model instead is motivated below: The model is almost as compact as the best one. This ensures its real-time implementability, which is important for the realization of high performance, model-based trajectory tracking control strategies. Normally, for their realization the most compact and system dynamics at its best reproducing model would be used. But since in the following both the efficiency of a parameter identification concept (cp. Chap. 6) and the influence of model inaccuracies on a feedback linearization of the system dynamics (cp. Chap. 8) will be studied, small derivations are 'useful', for (numerically) evaluating these concepts.

5.5 Inverse and direct dynamics of the SpiderMill

In the following chapters the reduced Newton-Euler model of the SpiderMill is used. The derived inverse dynamic model can be expressed in several forms, only using the active or also the passive coordinates. If only the active coordinates are used the model can be expressed by

$$\boldsymbol{\tau} = \boldsymbol{\tau}_{rb} + \boldsymbol{\tau}_f = \mathbf{M}\left(\mathbf{q}_a\right) \cdot \ddot{\mathbf{q}}_a + \mathbf{C}\left(\mathbf{q}_a, \dot{\mathbf{q}}_a\right) \cdot \dot{\mathbf{q}}_a + \mathbf{G}\left(\mathbf{q}_a\right) + \boldsymbol{\tau}_f\left(\dot{\mathbf{q}}_a\right), \qquad (5.24)$$

with $\boldsymbol{\tau}$ representing the torque vector of the overall model, $\boldsymbol{\tau}_{rb}$ the torque vector of the rigid body dynamics and $\boldsymbol{\tau}_f$ the vector of torques due to friction. But this model is too complex to be solved in real-time on the used dSPACE system. In order to have a model which can be calculated in real time and also be inverted (right dimension of $\mathbf{M}$) the following model containing active and passive joint coordinates is introduced for the SpiderMill

$$\boldsymbol{\tau} = \mathbf{M}\left(\mathbf{q}\right) \cdot \ddot{\mathbf{q}}_a + \mathbf{C}\left(\mathbf{q}\right) \cdot \dot{\mathbf{q}}^2 + \mathbf{G}\left(\mathbf{q}\right) + \boldsymbol{\tau}_f\left(\dot{\mathbf{q}}\right), \qquad (5.25)$$

where $\mathbf{q} = \left[\mathbf{q}_a^T, \ \mathbf{q}_p^T\right]^T$. From Eq. (5.25) the direct dynamics can be derived to

$$\ddot{\mathbf{q}}_a = \mathbf{M}^{-1} \cdot \left(\tau - \mathbf{C} \cdot \dot{\mathbf{q}}^2 - \mathbf{G} - \boldsymbol{\tau}_f\right), \qquad (5.26)$$

which can also be expressed in form of a state-space model:

states: $x_1 = q_{a1} = s_l$ inputs: $u_1 = \tau_l$ outputs: $y_1 = s_l$

    $x_2 = q_{a2} = s_r$ $u_2 = \tau_r$ $y_2 = s_r.$

    $x_3 = \dot{x}_1 = \dot{s}_l$

    $x_4 = \dot{x}_2 = \dot{s}_r$ (5.27)

model:

$$
\begin{bmatrix} \dot{x}_1 \\ \dot{x}_2 \\ \dot{x}_3 \\ \dot{x}_4 \end{bmatrix} = \begin{bmatrix} x_3 \\ x_4 \\ [-\mathbf{M}^{-1}(\mathbf{C}\dot{\mathbf{q}}^2 + \mathbf{G} + \boldsymbol{\tau}_f)]_{(1)} \\ [-\mathbf{M}^{-1}(\mathbf{C}\dot{\mathbf{q}}^2 + \mathbf{G} + \boldsymbol{\tau}_f)]_{(2)} \end{bmatrix} + \begin{bmatrix} 0 & 0 \\ 0 & 0 \\ [\mathbf{M}^{-1}]_{(1,1)} & [\mathbf{M}^{-1}]_{(1,2)} \\ [\mathbf{M}^{-1}]_{(2,1)} & [\mathbf{M}^{-1}]_{(2,2)} \end{bmatrix} \begin{bmatrix} u_1 \\ u_2 \end{bmatrix}
$$

$$
\begin{bmatrix} y_1 \\ y_2 \end{bmatrix} = \begin{bmatrix} x_1 \\ x_2 \end{bmatrix} \tag{5.28}
$$

where (i) refers to the ith row of the expression and (i,j) to the matrix element in row i and column j.

5.6 Simulation of mechatronic systems

For the design, analysis and simulation of mechatronic systems software tools are applied to an increasing degree. They are used for the modeling and analysis of individual components e.g. mechanical parts or electrical devices up to the entire system including the control structure. In order to do so, two kinds of strategies have to be distinguished. One way is to model all components in one software tool the other one is to couple two or more tools, with the inherent advantage, that each component or structure, respectively, can be implemented in the best suited tool. In order to make the choice in case of the SpiderMill the two – at the Institute of Control Systems – available software-tools as well as their possibile coupling are discussed below.

5.6.1 Theoretical background of the applied MBS tools

The theory of this section is mainly taken from [WK03, Sch03, SW04].

Relative and absolute coordinates

The structure of the equations of motions depends largely on the choice of the coordinates. In case of MSC.ADAMS absolute coordinates are applied, leading to a model with a large number of configuration variables (large dimension of $\mathbf{q}$) and a corresponding large number of simple(r) constraint equations. As a consequence $\mathbf{M}(\mathbf{q})$ is particularly simple and of block diagonal structure [WK03]. The uniformity of the equations and the simplicity of the constraint equations speak for this coordinate choice and are the reason, why most commercial software packages use absolute coordinates. The relative coordinate approach, used in SimMechanics, in contrast minimizes the number of necessary coordinates for representing the configuration by implicitly parameterizing certain constraints between the individual bodies of a mechanism. As a result the dimension of $\mathbf{q}$ and the number of constraint equations decrease. The main drawbacks of this approach are the dense mass matrix $\mathbf{M}(\mathbf{q})$, which contains the, now more complicate, constraints implicitly [Sch03]. The higher computation costs for constructing and inverting $\mathbf{M}(\mathbf{q})$ contributes significantly to the overall computational costs. But with recursive computational techniques it is possible to factorize the mass matrix and to invert it much more efficiently.

Theoretical background of MSC.ADAMS

In order to set up the equations of motion in MSC.ADAMS the Lagrangian equations of the first type are used in the following form [Ram00, SW04, NH01]

$$\frac{d}{dt}\left(\frac{\partial \mathcal{L}}{\partial \dot{q}_i}\right) - \frac{\partial \mathcal{L}}{\partial q_i} + \sum_{j=1}^{N_c} \lambda_j \frac{\partial \Gamma_j}{\partial q_i} - \sum_{k=1}^{l} \boldsymbol{\tau}_k^{ext} \frac{\partial \mathbf{r}_k}{\partial q_i} = 0, \quad i = 1, \dots, N_q \qquad (5.29\text{a})$$

$$\text{with} \quad \Gamma_j(\mathbf{q}, t) = 0, \quad j = 1, \dots, N_c \qquad (5.29\text{b})$$

its parameters are explained in the context of Eq. (B.8) in detail, only the external forces $\boldsymbol{\tau}_k^{ext}$ with their force application points $\mathbf{r}_k$ are concretized. As a consequence of the redundant (not independent) coordinates q_i a simpler expressions for $\mathcal{L}$ is obtained. These q_is are subject to the constraints Γ_j introduced by the joints of a mechanism. Based on Eq. (5.29) a set of nonlinear differential-algebraic equations (DAE) of second order and index-3 is obtained [SW04].

Theoretical background of SimMechanics

In SimMechanics the equations of motion are expressed in form of [WK03, Sch03]

$$\mathbf{M}(\mathbf{q})\ddot{\mathbf{x}} = \mathbf{f}(t, \mathbf{q}, \dot{\mathbf{x}}) + \tilde{\mathbf{J}}^T(\mathbf{q}) \left(\frac{\partial \Gamma(\mathbf{q}, t)}{\partial q}\right)^T \boldsymbol{\lambda} \qquad (5.30\text{a})$$

$$\text{with} \quad \dot{\mathbf{q}} = \tilde{\mathbf{J}}(\mathbf{q})\dot{\mathbf{x}} \quad \text{and} \quad \Gamma(\mathbf{q}, t) = 0, \qquad (5.30\text{b})$$

where $\tilde{\mathbf{J}}$ is the Jacobi matrix and $\mathbf{x}$ is the generalized position. The mass matrix $\mathbf{M}$ is symmetric positive-definite and $\mathbf{f}$ represents the contribution of the centrifugal, Coriolis, and external forces.

In SimMechanics closed-loops structures are reduced to open-loop topologies by cutting joints, which become replaced by a set of constraint equations. Thus it is ensured, that the new system behaves like the original one. Naturally, the structure of the resulting equations depends strongly on the choice of the cutting joint. Also in case of SimMechanics a index-3 DAE problem has to be solved. But Simulink cannot solve higher-index DAEs – it can only be used to model systems governed by ordinary differential equations (ODEs) or index-1 DAEs [SR97, SRK99]. Therefore the system description is transformed into ODEs in order to solve them with the Simulink ODE solver suite.

5.6.2 Combined or stand-alone simulation

Several possibilities for the coupling of two or more simulation tools exist, whereas all of it have both advantages and disadvantages [HD00, KGLS07a]. In this section the goal is the development of an appropriate simulation environment for the testing of control algorithms in the context of parallel robots. First the often applied coupling of MSC.ADAMS and Matlab/Simulink in form of a co-simulation will be shortly discussed: This kind of coupling is the only effective one, since it allows more than an implementation of only linear control concepts, as the most of the other concepts do. But its suffers from long simulation times and simulation faults or even instability if not an appropriate solver (integrator) combination is chosen in case of the discrete mode. Examples for successful applications of co-simulation are amongst others [DFM05, Rit05, KGLS07a].

One aspect has to be pointed out in oder to understand the problem better. For the modeling and simulation of the dynamics of a mechanical system – for which the relative locations or the movements of the individual components to each other, respectively, plays an important role – the usage of some kind of MBS system is unavoidable. In contrast, electronic (sub-)systems e.g. drives can be implemented more effective in systems like Matlab/Simulink. This directly implies the following main question: Can a MBS extension toolbox like SimMechanics be used to effectively simulate MBS systems in case of parallel robots? If the answer is yes, the disadvantages of a co-simulation can be avoided, implementing the control algorithms, the sensors, the actuators and the mechanical structure in one stand-alone simulation. Beyond all question a powerful MBS tool like MSC.ADAMS simulates complex dynamical systems better than an extension toolbox. But since here the goal is to develop an effective simulation environment for the parallel robot Spider-Mill, which represents rigid body dynamics sufficient enough, smaller inaccuracies are no problem. In order to analyze the modeling possibilities of SimMechanics in more detail, the model of the SpiderMill introduced in Sec. 2.2.5 is compared to the MSC.ADAMS model. The simulated forces are compared for the circle trajectories planned for a jerk limit of $1.6 \, \mathrm{Nm \, s^{-1}}$. The direct comparison of both sets of forces are shown in Fig. 5.12 together with the absolute and relative errors of the forces. The plots show that the principle results are the same for both tools.

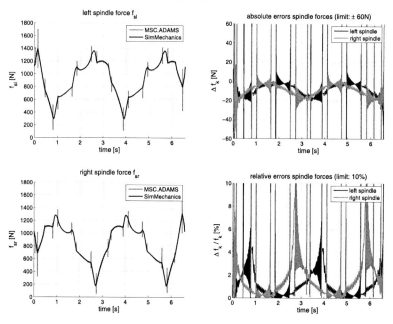

Figure 5.12: Comparison: MSC.ADAMS ↔ SimMechanics

The outliers in case of MSC.ADAMS can among other things be explained by an improper solver choice (here: default). At the physical demonstrator no such outliers of the forces can be observed. In order to compare both sets of forces without their influence, a fifth order Butterworth noise filter with a cut-off frequency of $f_{cut} = 15 \, \mathrm{Hz}$ is applied to the MSC.ADAMS values. Due to the offline application of the filter a phase shift can be avoided. The shown results for the comparison in case of the SpiderMill justifies the use of only one simulation tool. For other mechanical system this maybe does not hold. Therefore the choice of an adequate simulation environment is a first and very important step analyzing mechanical

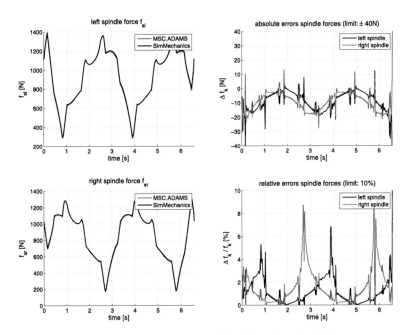

Figure 5.13: Comparison: MSC.ADAMS (filtered) $\leftrightarrow$ SimMechanics

structures.

Nevertheless here both models of the SpiderMill will be employed. For the parameter-identification in Chap. 6 the MSC.ADAMS model alone is used, whereas in case of the test of observers and control algorithms in Chap. 9 and 10 the SimMechanics model is the better choice.

5.7 Conclusions and future works

In this chapter dynamic models for the SpiderMill have been derived based on the structure(s) achieved in Chap. 4 by applying the proposed simplification strategy. The results obtained by using several standard modeling approaches confirm the assumptions of Chap. 4 and lead to well suited models for performing the following steps. Beyond the model used in Chap. 6 for identifying the rigid body parameters and friction terms of the SpiderMill also reduced models for implementing standard (decentralized) control concepts in Chap. 10 have been derived. A detailed analysis of two MBS simulation tools – MSC.ADAMS and the Matlab/Simulink SimMechanics toolbox – justify the usage of both in the following steps. With this chapter the dynamic modeling of the SpiderMill is completed, with respect to applying different modeling approaches. Following interesting questions are related to a further reduction of the model complexity as well as having a deeper look on the different dynamic coupling terms resulting from the individual approaches.

6 Parameter identification

The importance of accurate dynamic models – with well known dynamic parameters – in case of model-based trajectory tracking control and the compensation of friction [BI05, dWL97, Tom00, ZP02] for improving the tracking accuracy are in robotics often discussed points. In case of this thesis they are mainly analyzed in Chap. 10. In this context parameter identification plays an important role. For most robots the rigid body parameters cannot, even with very high effort, be calculated exactly. Especially in case of the friction parameters the application of a parameter identification strategy is unavoidable.

In case of serial robots each link can be moved independently from the others. Consequently, trajectories for each link, only exciting its parameters and reducing the influence of other links, can be planned. Starting at the end-effector the parameters of each link can be estimated separately in a recursive manner. In contrast, in case of parallel robots all links can only be moved together. Therefore all parameters must be collectively identified. This complicates the trajectory planning and leads also to a higher computational effort.

Normally two different strategies are applied in case of parallel robots: the *indirect identification method* [GHA04a] and the *direct identification method* [ABHG04, AAH06]. They have in common, that they use special planned trajectories for the end-effector motion that optimally excite the system. In case of the indirect identification a separate estimation of the rigid body and friction parameters is performed. Both of its parts consists of two steps. First local models are identified, which are used in a second step for the actual parameter estimation. Its main disadvantage is the effort for identifying several local models. In case of the direct parameter identification an identification of local models is not necessary (one step strategy). The whole identification bases on one trajectory, which excites the system optimally. The concept can be applied in several ways: for a joint or separated estimation of the rigid body and friction parameters.

A further special aspect has to be considered in case of parallel robots. Most identification strategies only take the friction losses in the active joints into account. But in [AAH06] it has been clearly demonstrated, that for parallel robots with their (many) passive joints also their friction losses play a very important role and have to be taken into account in order to improve the dynamic models.

Having investigated both basic strategies[1] the following procedure – combining them – for estimating the parameters of the inverse dynamic model seems applicable (confirmed by the results below) to the SpiderMill [KML08a, KML08b, KML09][2]:

- Indirect identification of the friction parameters (considering passive joint friction)

- Direct identification of the rigid body parameters based on the knowledge of the friction parameters

Planning the required motions boundary conditions and kinematic constraints, e.g. the available workspace of the manipulator or the limitations of actuator velocities and accelerations are considered.

This chapter is structured as follows: The model basis for the rigid body and friction parameter identification is introduced in Sec. 6.1. In Sec. 6.2 the applied procedure for the friction parameter identification is explained and in Sec. 6.3 that for the rigid body parameters. The reasons for the proposed combination of the two identification strategies are given in Sec. 6.4. In Sec. 6.5 simulated and experimental results applying the identification concept to the MBS model of the SpiderMill and to the physical demonstrator are summarized (also discussing further aspects). Conclusions are drawn in Sec. 6.6.

6.1 The inverse dynamic model as basis for parameter identification

6.1.1 Rigid body model and (passive) joint friction

Applying a standard dynamic modeling approach, cp. Chap. 5, to the simplified structure (of the SpiderMill), always leads to a parameter-linear model [KG85, KK87, KD02].

$$\boldsymbol{\tau}_{rb} = \mathbf{H}_{rb}\left(\mathbf{q}, \dot{\mathbf{q}}, \ddot{\mathbf{q}}\right) \cdot \mathbf{p}_{rb} \qquad \text{with} \quad \mathbf{q} = [\mathbf{q}_a, \mathbf{q}_p], \qquad (6.1)$$

where $\boldsymbol{\tau}_{rb}$ is the torque vector of the rigid body dynamics (cp. also Sec. 5.5), $\mathbf{H}_{rb}$ the regressor matrix only containing known kinematic variables (e.q. lengths or angles) and $\mathbf{p}_{rb}$ the vector of the dynamic parameters of the rigid body model. The simplified model of the SpiderMill contains both active and passive joints

---

[1]The motivation for it is given in Sec. 6.4 after the basic theory is introduced.

[2]The results presented in the following are taken (cited) from [KML09]. Wherever necessary extensions have been made.

within the limbs of the simplified structure. Therefore the friction behavior of the passive joints and the friction losses in the drive systems have to be considered. Regardless of the applied approach for dynamic modeling both friction parts can be expressed as an additive extension of the parameter-linear model

$$\boldsymbol{\tau} = \boldsymbol{\tau}_{rb} + \boldsymbol{\tau}_f \tag{6.2}$$

Dividing the friction terms in a passive and an active part Eq. (6.3) is obtained.

$$\boldsymbol{\tau} = \boldsymbol{\tau}_{rb} + \boldsymbol{\tau}_f = \mathbf{H}_{rb} \cdot \mathbf{p}_{rb} + \mathbf{D}_{pf} \cdot \mathbf{n}_{pf} + \mathbf{D}_{af} \cdot \mathbf{n}_{af} \tag{6.3}$$

with the matrix $\mathbf{D}_{pf}$ and $\mathbf{D}_{af}$ solely depending on geometric parameters (joint angles and lengths) and the vectors $\mathbf{n}_{pf}$ and $\mathbf{n}_{af}$ involving the friction torques of the passive joints and that of the actuation parts.
In general, friction can be modeled physically motivated e.g. by a combination of viscous and dry friction or phenomenologically by choosing appropriate trial functions reproducing the measured friction characteristics – cp. Sec. 6.1.2. The proper choice depends on the identification strategy.

6.1.2 Inverse dynamic model for parameter identification

In order to use the reduced Newton-Euler model of Sec. 5.1.3 for parameter identification the balances of torques, e.g. Eq. (5.8b), have to be extended by passive joint friction. Following the modeling approach of Chap. 4 the friction terms of the passive joints can be combined in the passive joints of the equivalent model J_{1l}, J_{mC1l}, J_{mM}, J_{mC1r} and J_{1r} – cp. Fig. 4.11. All of the substitution joints (J_i and J_{mi}) will be regarded as passive when a friction model is introduced. Like in Sec. 5.1.3 the procedure will be exemplarily shown for substitution part S_{1l}. Due to the fact that SpiderMill only consists of revolute joints the related joint frictions are modeled as friction torques $\mathbf{n}_i$ like sketched in Fig. 6.1.

$$
\begin{aligned}
&{}^B\mathbf{n}_{pal} - {}^B\mathbf{n}_{p1l} + {}^B\mathbf{r}_{Al} \times {}^B\mathbf{f}_{sl} + {}^B\mathbf{r}_{m1l} \times (0 \quad -m_1 g \quad 0)^T + {}^B\mathbf{r}_{1l} \times (0 \quad -m_2 g \quad 0)^T \\
&+ {}^B\mathbf{r}_{1l} \times (-{}^B\mathbf{f}_{1l}) = m_1 {}^B\mathbf{r}_{m1l} \times {}^B\dot{\mathbf{v}}_{m1l} + m_2 {}^B\mathbf{r}_{1l} \times {}^B\dot{\mathbf{v}}_{m2l} + {}^B\mathbf{I}_1^{c1} {}^B\dot{\boldsymbol{\omega}}_{1l} \\
&+ {}^B\boldsymbol{\omega}_{1l} \times ({}^B\mathbf{I}_{1l}^{m1l} {}^B\boldsymbol{\omega}_{1l}).
\end{aligned}
\tag{6.4}
$$

The balances of forces (here: Eq. (5.8a)) remain valid. From the general balance equations – now also including joint friction – solved for the spindle forces the following analytical model for the inverse dynamics is obtained

$$\boldsymbol{\tau} = \begin{bmatrix} \mathbf{f}_{sl}, & \mathbf{f}_{sr} \end{bmatrix}^T = \boldsymbol{\tau}_{rb} + \boldsymbol{\tau}_{jf} = \mathbf{H}_{rb} \cdot \mathbf{p}_{rb} + \mathbf{D}_{pf} \cdot \mathbf{n}_{pf}. \tag{6.5}$$

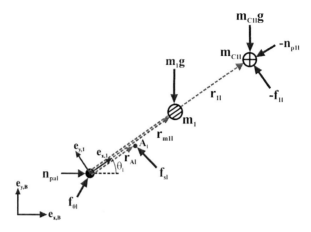

Figure 6.1: Forces, moments and friction moments acting on S_{1l} (and S_{2l})

with $\mathbf{D}_{pf}$ transforming the joint friction torques into spindle forces. The vector of the dynamic parameters of the rigid body model $\mathbf{p}_{rb}$ contains eleven elements

$$
\begin{array}{llll}
p_1 = m_1\, d_{m1} & p_4 = m_3 & p_7 = m_4 & p_{10} = I_{m1,3} \\
p_2 = m_1\, d_{m1}^2 & p_5 = m_3\, d_{m3} & p_8 = m_4\, d_{m3} & p_{11} = I_{m3,3}. \\
p_3 = m_2 & p_6 = m_3\, d_{m3}^2 & p_9 = m_4\, d_{m3}^2 &
\end{array} \tag{6.6}
$$

The vector $\mathbf{n}_{pf}$ of the friction moments of the five passive joints of the simplified model is given by

$$
\mathbf{n}_{pf} = \begin{bmatrix} n_{pal}, & n_{p1l}, & n_{pM}, & n_{p1r}, & n_{par} \end{bmatrix}^T. \tag{6.7}
$$

In the next step the friction losses in the drive systems are considered in order to complete the inverse dynamic model (6.3). Due to the actuation principle of the robot the vector $\mathbf{n}_{af}$ consists of the friction moments of the two actuation parts

$$
\mathbf{n}_{af} = \begin{bmatrix} n_{al}, & n_{ar} \end{bmatrix}^T. \tag{6.8}
$$

The matrices $\mathbf{D}_{af}$ and $\mathbf{D}_{pf}$ of Eq. (6.3) transform the friction moments into forces. Based on [Dau01], characteristic curves of ball bearings are taken for the approximative static friction model. It only considers a Coulomb friction influence for the passive joints n_{pi} and the joints of the actuation systems n_{ai}

$$
\begin{aligned}
n_{pi} &= r_{pi} \cdot \operatorname{sgn}\left(\dot{q}_i\right) & i &\in \{al,\, 1l,\, M,\, 1r,\, ar\} \\
n_{ai} &= r_{ai} \cdot \operatorname{sgn}\left(\dot{q}_i\right) & i &\in \{l,\, r\},
\end{aligned} \tag{6.9}
$$

where r_{pi} and r_{ai} are called friction coefficients. Other possible models for friction amongst others including viscous damping can be found e.g. in [BI05, AAH06]. The main advantage of the standard model in Eq. (6.9) is its simpleness and its parameter linearity. Hence, the friction part in Eq. (6.3) can be transformed into the following parameter-linear form

$$\boldsymbol{\tau} = \boldsymbol{\tau}_{rb} + \boldsymbol{\tau}_{jf} + \boldsymbol{\tau}_{af} = \mathbf{H}_{rb} \cdot \mathbf{p}_{rb} + \mathbf{H}_f \cdot \mathbf{p}_f, \qquad (6.10)$$

where $\mathbf{p}_f$ contains the friction coefficients of the passive joints and the drive systems. The obtained inverse dynamic model is parameter-linear but not necessarily parameter-minimal. To get a parameter-minimal one, required for applying linear estimators, some further – below explained calculation steps – are necessary.

6.1.3 Parameter-minimal model

The parameter-minimality of the used inverse dynamic model is very important in case of parameter identification. For that reason, first the rigid body part of Eq. (6.10) is considered. If $\boldsymbol{H}_{rb}$ is of full-rank all elements of the parameter vector $\boldsymbol{p}_{rb}$ can be identified independent of each other. This form of the inverse dynamic model is said to be parameter-minimal.

For determining the minimal parameter set $\mathbf{p}_{rb,min}$, parameters $p_{rb,j}$ that do not influence the inverse dynamics (corresponding columns $\mathbf{h}_{rb,j}$ of $\mathbf{H}_{rb}$ are zero) have to be eliminated [GK86, GK90, GH00] from $\mathbf{p}_{rb}$ (and the corresponding columns from $\mathbf{H}_{rb}$). If the resulting matrix from $\boldsymbol{H}_{rb}$ is still not full-rank, linear dependencies of its columns have to be examined

$$\boldsymbol{h}_{rb,j} = \sum_{k=1}^{n_j} a_k^j \cdot \boldsymbol{h}_{rb,k}^j , \qquad (6.11)$$

with constant linear factors a_k^j. Parameters $p_{rb,j}$'s and associated columns $\boldsymbol{h}_{rb,j}$'s can be eliminated if the remaining parameters are replaced by

$$p_{rb,k,new}^j = p_{rb,k}^j + a_k^j \cdot p_{rb,j} \qquad k = 1 \ \dots \ n_j. \qquad (6.12)$$

The resulting parameter-linear form of the rigid body part

$$\boldsymbol{\tau} = \mathbf{H}_{rb}(\mathbf{q}_a, \dot{\mathbf{q}}_a, \ddot{\mathbf{q}}_a) \cdot \mathbf{p}_{rb}, \qquad (6.13)$$

has to be examined for minimality in its parameters by using a sequence of samples $(\mathbf{q}_a, \dot{\mathbf{q}}_a, \ddot{\mathbf{q}}_a)_{(i)}$, $i = 1 \dots N$ and calculating the rank of the matrix [Gau90]

$$\mathbf{W} = \begin{bmatrix} \mathbf{H}_{rb}(\mathbf{q}_a, \dot{\mathbf{q}}_a, \ddot{\mathbf{q}}_a)_{(1)} \\ \vdots \\ \mathbf{H}_{rb}(\mathbf{q}_a, \dot{\mathbf{q}}_a, \ddot{\mathbf{q}}_a)_{(N)} \end{bmatrix} . \qquad (6.14)$$

If it is of full rank, the rigid body part is parameter-minimal. Otherwise further reduction strategies, like e.q. the QR decomposition with pivoting [Gau90] have to be applied, in order to get it parameter-minimal

$$\boldsymbol{\tau}_{rb} = \boldsymbol{H}_{rb,min}\left(\boldsymbol{q}_a, \dot{\boldsymbol{q}}_a, \ddot{\boldsymbol{q}}_a\right) \cdot \boldsymbol{p}_{rb,min}. \tag{6.15}$$

For the derivation of the parameter-minimal form of the friction part of Eq. (6.10) in principle the same procedure like above can be applied. But in general the matrix $\boldsymbol{H}_f$ has no zero columns and also no linear dependencies occur. For that reason the parameter-linear form of the friction part is normally also its parameter-minimal form

$$\boldsymbol{\tau}_f = \boldsymbol{H}_{f,min}\left(\boldsymbol{q}_a, \dot{\boldsymbol{q}}_a\right) \cdot \boldsymbol{p}_{f,min}. \tag{6.16}$$

The combination of both parts yields the complete parameter-minimal model

$$\boldsymbol{\tau} = \boldsymbol{\tau}_{rb} + \boldsymbol{\tau}_f = (\ \underbrace{\boldsymbol{H}_{rb,min} \cdot \boldsymbol{p}_{rb,min}}_{\text{rigid body model}} + \underbrace{\boldsymbol{H}_{f,min} \cdot \boldsymbol{p}_{f,min}}_{\text{friction model}}\ ). \tag{6.17}$$

6.2 Friction identification

First the estimation of friction parameters of Eq. (6.17) is dealt with, before in a second step the rigid body parameters will be identified. In the following the two step identification procedure of the indirect identification method is explained. Applying the concept an optimal excitation of the system for friction parameter identification can be realized.

6.2.1 Identification of local friction models (Step 1)

In a first step local actuation friction models – assuming that the friction of the whole robot is concentrated in the actuators – are identified for different working points (WP) $\boldsymbol{x}_{WPj}$. For this purpose an excitation of the system dynamics is required, which has a strong influence on the friction and a low on the rigid body effects. Therefore, an end-effector trajectory passing the WPs with constant velocities ($\ddot{\boldsymbol{x}} = \boldsymbol{0}$) is used to reduce the influence of the inertia forces. As proposed in [GHA04a] the following relationship is assumed for the i-th active joint

$$\tau_{ijk}\left(\dot{q}_{a,ijk}\ ,\ \boldsymbol{x}_{WPj}\right) = \begin{bmatrix} 1 & \dot{q}_{a,ijk}^2 & \dot{q}_{a,ijk} & \mathrm{sgn}\left(\dot{q}_{a,ijk}\right) \end{bmatrix} \underbrace{\begin{bmatrix} \tilde{G}_i\left(\boldsymbol{x}_{WPj}\right) \\ \tilde{M}_{ii}\left(\boldsymbol{x}_{WPj}\right) \\ r_{i1}\left(\boldsymbol{x}_{WPj}\right) \\ r_{i2}\left(\boldsymbol{x}_{WPj}\right) \end{bmatrix}}_{\boldsymbol{p}_{f,act,ij}}, \tag{6.18}$$

where $\tilde{G}_i\,(\boldsymbol{x}_{WPj})$ is the constant gravitational influence, $\tilde{M}_{ii}\,(\boldsymbol{x}_{WPj})$ the resulting influence of inertia and $r_{i1}\,(\boldsymbol{x}_{WPj})$ and $r_{i2}\,(\boldsymbol{x}_{WPj})$ the local friction parameters. Summarizing the measurements for different velocities $\dot{q}_{a,ijk}$, $k = 1 \ldots N$ for the i-th actuator at working point $\boldsymbol{x}_{WPj}$ yields

$$
\underbrace{\begin{bmatrix} \tau_{ij1} \\ \vdots \\ \tau_{ijN} \end{bmatrix}}_{\boldsymbol{\Gamma}_{f,act,ij}} - \underbrace{\begin{bmatrix} 1 & \dot{q}^2_{a,ij1} & \dot{q}_{a,ij1} & \mathrm{sgn}\,(\dot{q}_{a,ij1}) \\ & & \vdots & \\ 1 & \dot{q}^2_{a,ijN} & \dot{q}_{a,ijN} & \mathrm{sgn}\,(\dot{q}_{a,ijN}) \end{bmatrix}}_{\boldsymbol{\Psi}_{f,act,ij}} \cdot \boldsymbol{p}_{f,act,ij} = \underbrace{\begin{bmatrix} e_{ij1} \\ \vdots \\ e_{ijN} \end{bmatrix}}_{\boldsymbol{\eta}_{f,act,ij}}, \tag{6.19}
$$

with the measurement vector $\boldsymbol{\Gamma}_{f,act,ij}$, the information matrix $\boldsymbol{\Psi}_{f,act,ij}$ and the error vector $\boldsymbol{\eta}_{f,act,ij}$. By applying a least squares estimator (LS-estimator) the parameter vector is determined by

$$
\boldsymbol{p}_{LS,f,act,ij} = \left(\boldsymbol{\Psi}^T_{f,act,ij} \cdot \boldsymbol{\Psi}_{f,act,ij}\right)^{-1} \cdot \boldsymbol{\Psi}^T_{f,act,ij} \cdot \boldsymbol{\Gamma}_{f,act,ij}. \tag{6.20}
$$

This procedure is repeated for all chosen WPs.

6.2.2 Calculation of joint friction parameter sets from the local models (Step 2)

The actuation friction $\tau_{f,ijk}$ at the local working point $\boldsymbol{x}_{WPj}$ can be determined using the parameters $r_{i1}(\boldsymbol{x}_{WPj})$ and $r_{i2}(\boldsymbol{x}_{WPj})$ determined above

$$
\tau_{f,ijk} = r_{i1}(\boldsymbol{x}_{WPj})\,\dot{q}_{a,ijk} + r_{i2}(\boldsymbol{x}_{WPj})\,\mathrm{sgn}\,(\dot{q}_{a,ijk})\,. \tag{6.21}
$$

Combining the results at all WPs, an overall measurement equation can be stated

$$
\underbrace{\begin{bmatrix} \tau_{f,111} \\ \vdots \\ \tau_{f,LMN} \end{bmatrix}}_{\boldsymbol{\Gamma}_f} - \underbrace{\begin{bmatrix} \boldsymbol{H}_{f,min,11}\,(\boldsymbol{q}_{a,11}, \dot{\boldsymbol{q}}_{a,11}) \\ \vdots \\ \boldsymbol{H}_{f,min,LN}\,(\boldsymbol{q}_{a,LN}, \dot{\boldsymbol{q}}_{a,LN}) \end{bmatrix}}_{\boldsymbol{\Psi}_f} \cdot \boldsymbol{p}_{f,min} = \underbrace{\begin{bmatrix} e_{ij1} \\ \vdots \\ e_{ijN} \end{bmatrix}}_{\boldsymbol{\eta}_f}, \tag{6.22}
$$

where $\boldsymbol{\eta}_f$ is the error vector. Applying the LS-estimator the parameter vector is determined by

$$
\boldsymbol{p}_{LS,f,min} = \left(\boldsymbol{\Psi}^T_f \cdot \boldsymbol{\Psi}_f\right)^{-1} \cdot \boldsymbol{\Psi}^T_f \cdot \boldsymbol{\Gamma}_f. \tag{6.23}
$$

This parameter vector is used as $\boldsymbol{p}_{f,min}$ in Eq. (6.17).

Ideally, the WPs $\boldsymbol{x}_{WPj}$ should be defined in a way that the friction parameters of all joints are excited optimally and equally. Here, they are selected symmetrically on a circle in the workspace, as proposed in [GHA04a].

6.3 Rigid Body Identification (one step strategy)

In the following the problem of identifying the rigid body parameters is discussed. For this purpose the direct identification method (one step strategy) is applied. A trajectory optimally exciting the system for rigid body parameter identification is planned for the active joints $\boldsymbol{q}_a$. The trajectory and the generalized forces $\boldsymbol{\tau}$ are sampled at N equidistant time instants. Collecting all sampled data by subtracting the identified frictions determined in Sec. 6.2 yields

$$\underbrace{\begin{bmatrix} \boldsymbol{\tau}_1 - \boldsymbol{H}_{f,min,1}\left(\boldsymbol{q}_a,\dot{\boldsymbol{q}}_a\right)\cdot\boldsymbol{p}_{f,min} \\ \vdots \\ \boldsymbol{\tau}_N - \boldsymbol{H}_{f,min,N}\left(\boldsymbol{q}_a,\dot{\boldsymbol{q}}_a\right)\cdot\boldsymbol{p}_{f,min} \end{bmatrix}}_{\boldsymbol{\Gamma}_{rb}} - \underbrace{\begin{bmatrix} \boldsymbol{H}_{rb,min,1}\left(\boldsymbol{q}_{a1},\dot{\boldsymbol{q}}_{a1},\ddot{\boldsymbol{q}}_{a1}\right) \\ \vdots \\ \boldsymbol{H}_{rb,min,N}\left(\boldsymbol{q}_{aN},\dot{\boldsymbol{q}}_{aN},\ddot{\boldsymbol{q}}_{aN}\right) \end{bmatrix}}_{\boldsymbol{\Psi}_{rb}} \cdot \boldsymbol{p}_{rb,min} = \underbrace{\begin{bmatrix} e_{ij1} \\ \vdots \\ e_{ijN} \end{bmatrix}}_{\boldsymbol{\eta}_{rb}},$$

$$(6.24)$$

with the error vector $\boldsymbol{\eta}_{rb}$. Applying the LS-estimation the parameter vector is determined to

$$\boldsymbol{p}_{LS,rb,min} = \left(\boldsymbol{\Psi}_{rb}^T\cdot\boldsymbol{\Psi}_{rb}\right)^{-1}\cdot\boldsymbol{\Psi}_{rb}^T\cdot\boldsymbol{\Gamma}_{rb}. \qquad (6.25)$$

In order to evaluate the quality of the parameter estimation the relative error between the estimated $\boldsymbol{p}_{LS,rb,min}$ and real parameter vector $\boldsymbol{p}_{rb,min}$ is examined[3]

$$\frac{\left\|\boldsymbol{p}_{LS,rb,min}-\boldsymbol{p}_{rb,min}\right\|}{\left\|\boldsymbol{p}_{rb,min}\right\|} \le \underbrace{\left\|\boldsymbol{\Psi}_{rb}\right\|\cdot\left\|\left(\boldsymbol{\Psi}_{rb}^T\cdot\boldsymbol{\Psi}_{rb}\right)^{-1}\cdot\boldsymbol{\Psi}_{rb}^T\right\|}_{=\kappa(\boldsymbol{\Psi}_{rb})}\cdot\frac{\left\|\boldsymbol{\eta}_{rb}\right\|}{\left\|\boldsymbol{\Gamma}_{rb}\right\|}, \qquad (6.26)$$

where $\kappa\left(\boldsymbol{\Psi}_{rb}\right)$ is the condition number and $\|\cdot\|$ the spectral norm [BIG74]. It describes the ratio of the biggest and the smallest singular value of the information matrix $\boldsymbol{\Psi}_{rb}$. Its value, varying from $1 \dots \infty$, denotes the uncertainty of the estimation and can be used to evaluate the quality of the excitation of the system. A minimal $\kappa\left(\boldsymbol{\Psi}_{rb}\right)$ leads to an uniform excitation of all elements of $\boldsymbol{p}_{rb,min}$. Since the value of the condition number is determined by the active joint variables $\boldsymbol{q}_a$ and their derivatives $\dot{\boldsymbol{q}}_a$ and $\ddot{\boldsymbol{q}}_a$, it can also be used for optimizing the trajectory generation. Therefor a parameterizable basis function

$$\boldsymbol{q}_a = \boldsymbol{q}_a\left(t,\boldsymbol{\epsilon}\right), \qquad (6.27)$$

with $\boldsymbol{\epsilon}$ containing the parameters of the trajectories is used. Taking into account the constraints for the workspace as well as that for the velocities and acceleration

---

[3]Eq. (6.26) defines an upper bound for the relative estimation error [DA98]. For determining the related condition number the knowledge of the real parameter vector is not necessary.

of the active joints, the optimization problem for trajectory generation can be formulated as

$$\text{cost function:} \quad \min_{\boldsymbol{\epsilon}} \, \kappa \left(\boldsymbol{\Psi}_{rb} \right)$$

$$\text{parametrized function:} \quad \boldsymbol{q}_a = \boldsymbol{q}_a \left(t, \boldsymbol{\epsilon} \right) \quad \boldsymbol{q}_a \in \mathbb{R}^n \quad t \in [t_0, t_e]$$

$$\text{constraints}^4 \!: \quad \boldsymbol{x}_{min} \leq \boldsymbol{x} \leq \boldsymbol{x}_{max}$$

$$\dot{\boldsymbol{q}}_{a,min} \leq \dot{\boldsymbol{q}}_a \left(t_k, \boldsymbol{\epsilon} \right) \leq \dot{\boldsymbol{q}}_{a,max}$$

$$\boldsymbol{\tau}_{min} \leq \boldsymbol{\tau} \left(\boldsymbol{q}_a, \dot{\boldsymbol{q}}_a, \ddot{\boldsymbol{q}}_a \right) \leq \boldsymbol{\tau}_{max}$$

$$t_k \in [t_0, t_e] \quad k = 1 \ldots N. \tag{6.28}$$

In order to achieve a practically sufficient dynamic excitation frequency domain basis functions are often suggested. Here a finite Fourier series for each active joint is applied

$$q_{a,i} \left(t, \boldsymbol{\epsilon}^i \right) = q_{a,i0} + \sum_{k=0}^{n_f} \left(\frac{\mu_k^i}{k \cdot \omega_f} \cdot \sin \left(k \cdot \omega_f \cdot t \right) - \frac{\nu_k^i}{k \cdot \omega_f} \cdot \cos \left(k \cdot \omega_f \cdot t \right) \right). \tag{6.29}$$

Thus, the parameter vector of the ith active joint $(i = 1 \ldots n)$ is given by

$$\boldsymbol{\epsilon} = \begin{bmatrix} \omega_f & \boldsymbol{\epsilon}^1 & \ldots & \boldsymbol{\epsilon}^n \end{bmatrix} \quad \text{with} \quad \boldsymbol{\epsilon}^i = \begin{bmatrix} q_{a,i0} & \mu_1^i & \ldots & \mu_{n_f}^i & \nu_1^i & \ldots & \nu_{n_f}^i \end{bmatrix}. \tag{6.30}$$

The number of elements of $\boldsymbol{\epsilon} = n \cdot (2 \cdot n_f + 1) + 1$ is the 'degree of freedom' of the optimization problem. As solutions of the problem (6.28) a minimal condition number $\kappa \left(\boldsymbol{\Psi}_{rb} \right)$ and the trajectories for the active joints are obtained.

Another strategies for finding suitable exciting trajectories can be found in [ACC99]. In contrast to the strategy above, beside a condition number additionally the magnitude of the minimum singular value is used there as part of the cost function. Furthermore an interpolation algorithm is implemented in order to provide smooth trajectories.

Beside the here proposed strategy for parameter-identification also other ones, explicitly determining which dynamic terms are candidates for identification or are not identifiable (and therefore can be discarded from the model), exist [ACC99, WSO02]. In contrast to the strategy proposed above and that in [ACC99] several trajectories of varying dynamics (different platform movements and velocities are

---

[4]Relation operators are only defined for vectors (or matrices) in combination with introduced norms. But since it is a common notation in robotics to apply them directly, they are also used in this thesis in an element-wise way.

used in order to enhance the dynamic dominance of individual terms) are applied in [WSO02] for identifying the individual parts of the model in multi (iterative) stages also considering the influence of friction. Other approaches [FMAPV08] use a similar identification procedure like the one above, but strongly emphasize the importance of physical feasibility in case of parameter identification, cp. Sec. 6.5.5.

6.4 Motivation for the identification procedure

After introducing the basic concepts and their theoretical background it is possible to explain in more detail, why the direct and indirect identification method are combined in that way in case of the SpiderMill.
Performing simulative studies, a much better (since lower) condition number has been achieved for trajectories planned based on the direct identification method in the workspace of the SpiderMill (cp. Sec. 6.5.2). Due to that reason the direct identification method is used for the identification of the rigid body parameters of the inverse dynamic model. But the friction influence cannot be neglected and would lead to an incorrect estimation. Since the trajectories, applied for rigid body identification do not optimally excite the system in case of friction (they have been determined based on a model without friction), an appropriate friction model has to be identified in advance. For that purpose the indirect identification method is applied. With the identified friction model of the first step it is possible to compensate its influence in case of rigid body parameter identification at the demonstrator.

6.5 Simulation and experimental results

In the following the simulation and experimental results, obtained by applying the above introduced identification concept to the MBS model of the SpiderMill and the physical demonstrator are summarized.

6.5.1 Parameter-minimal inverse dynamic model

Applying the method described in Sec. 6.1.3 to the rigid body part of the inverse dynamic model Eq. (6.10), the following minimal parameter set is obtained

$$p_{rb,min,1} = d_1 m_2 + d_1 m_3 + d_{c1} m_1 + \frac{d_1 d_{c3}}{2 d_3} m_4$$

$$p_{rb,min,2} = d_1^2 m_2 + d_1^2 m_3 + d_{c1}^2 m_1 + \frac{d_1^2 d_{c3}}{2 d_3} m_4 + I_{c1,2}$$

$$p_{rb,min,3} = d_{c3} m_3 + \frac{d_{c3}}{2} m_4$$

$$p_{rb,min,4} = m_4 - \frac{d_{c3}}{d_3} m_4$$

$$p_{rb,min,5} = 2 d_{c3}^2 m_3 + d_{c3}^2 m_4 + 2 I_{c3,2}. \tag{6.31}$$

The friction part of the model (6.10) is already in parameter-minimal form.

6.5.2 Trajectory planning

The trajectory optimally exciting the system for rigid body parameter identification can be determined by solving the optimization problem (6.28) using a parameterized Fourier series as trial function for the active joint trajectories $q_a(t) = [s_l(t) , s_r(t)]^T$. For calculating the actuator constraints, however, the parameter set of the inverse dynamic model is needed in advance. Since it is not available at that time, the calculated parameters set from Tab. A.2 is used instead. The optimization problem (6.28) is solved then using SNOPT[5].

Considering all constraints at $N = 36$ sampling instants a condition number of $\kappa = 28.4$ is achieved, if a Fourier series with length $n_f = 9$ and a period $T_f = 4.488$ s is used as identification trajectory for the active joints q_a (cp. Fig. 6.2).

Beside the identification trajectory a further, also highly dynamical verification trajectory (cp. Fig. 6.3) – also based on a finite Fourier series and considering the limitations – is planned in order to verify the following results of the identification strategy.

---

[5]SNOPT is a general-purpose software tool for constrained optimization. It minimizes a nonlinear cost function subject to constrained variables. SNOPT applies a sequential quadratic programming (SQP) method using limited-memory quasi-Newton approximations [GMS05].

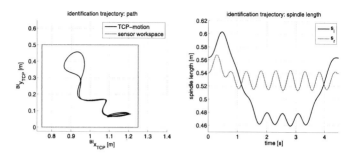

Figure 6.2: Identification trajectory with respect to TCP_A

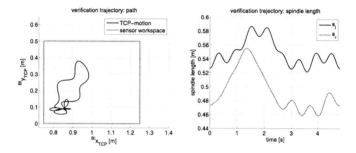

Figure 6.3: Verification trajectory with respect to TCP_A

parameter	$\mathbf{p}_{rb,min,calc}$	$\mathbf{p}_{rb,min,MBS}$	$\mathbf{p}_{rb,min,dem}$
$p_{rb,min,1}$	59.3028	64.7383	23.8458
$p_{rb,min,2}$	57.0641	63.2687	66.9369
$p_{rb,min,3}$	12.1140	16.7899	10.6589
$p_{rb,min,4}$	8.9450	-0.3692	-9.4155
$p_{rb,min,5}$	15.7119	30.2864	24.5169

Table 6.1: Numerical values for the minimal parameter vector $\mathbf{p}_{rb,min}$

In order to be able to verify the improvement in case of identified, instead of calculated parameters, the spindle forces of the simplified model, whose parameters

$p_{rb,min,calc}$ are calculated from CAD construction data (cp. Table A.2 and Table 6.1), are compared to that of the MSC.ADAMS model in Fig. 6.4. For friction

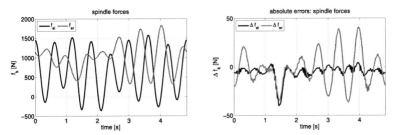

Figure 6.4: Inverse dynamics with calculated parameters ↔ MSC.ADAMS for the verification trajectory

identification 12 WPs located on a circle, as shown in Fig. 6.5, are chosen. In each of this WPs, motion trajectories passing the WPs with constant end-effector velocity – e.g. the straight line (tangent) for WP 8 – are used for identification.

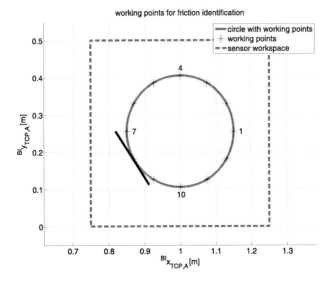

Figure 6.5: Working points for friction identification

6.5.3 Simulation results

The proposed identification procedure is first verified using MSC.ADAMS simulation. Since friction is not implemented in the model, only the rigid body parameters can be compared. Using the trajectories described in Sec. 6.5.2 the rigid body parameters vector $\boldsymbol{p}_{rb,min}$ is determined using the LS-estimator Eq. (6.25).

To verify the identified inverse dynamic model (6.15) the simulated and the calculated actuator forces $\boldsymbol{\tau}$ are compared to each other for the verification trajectory. Obviously, the simplified dynamic model with the identified parameters excellently

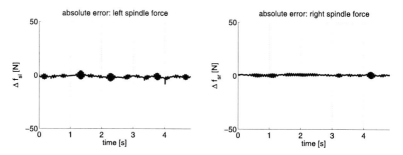

Figure 6.6: Inverse dynamics with estimated parameters $\leftrightarrow$ MSC.ADAMS for the verification trajectory

reproduces the behavior of the MSC.ADAMS model. Comparing Fig. 6.6 with Fig. 6.4 shows a significant improvement of the model using identified parameters $\mathbf{p}_{rb,min,MBS}$ (cp. Table 6.1) instead of the calculated ($\mathbf{p}_{rb,min,calc}$) ones.

6.5.4 Experimental results

The identification experiments at the demonstrator are carried out using the encoders for determining the active joint coordinates $\boldsymbol{q}_a$. The values of $\dot{\boldsymbol{q}}_a$ and $\ddot{\boldsymbol{q}}_a$ are calculated with Matlab applying difference quotients. The $\boldsymbol{\tau}$'s are determined indirectly by measuring the current of each actuator using LEM converters with a fifth order Butterworth noise filter (cut-off frequency: $f_{cut} = 6$ Hz). For the trajectory tracking standard servo controllers are applied.

For the friction identification local actuation friction models are identified for the 12 WPs located at the circle (cp. Fig. 6.5) for 5 different but constant end-effector velocities. The values of the $\boldsymbol{\tau}$'s, the $\boldsymbol{q}_a$'s and $\dot{\boldsymbol{q}}_a$'s are determined for these WPs. Based on the measurement vector $\boldsymbol{\Gamma}_{f,act,ij}$ and the information matrix $\boldsymbol{\psi}_{f,act,ij}$, cp.

Eq. (6.19), the parameter vector $\boldsymbol{p}_{f,act,ij}$ for each local actuation model i is identified using the LS-estimator, cp. Eq. (6.20). In Fig. 6.7 the measured forces are compared to the ones calculated using the identified models for both actuators at WP 8.

The joint friction models are separately determined as described in Sec. 6.2 and

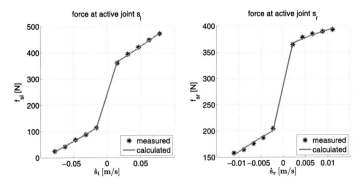

Figure 6.7: Actuation friction model for working point 8

presented in Fig. 6.8. Due to the fact, that the passive and the pseudo active joints of the SpiderMill are ball bearings, their friction characteristics, as well as that of the active joints can be approximately modeled by dry friction [Dau01]

$$n = r_i \cdot \text{sign}(\alpha). \tag{6.32}$$

With Eq. (6.32) as basis function the following friction models have been identified for the active

$$n_{al} = 0.0520 \cdot \text{sign}\left(\dot{s}_l\right), \qquad n_{ar} = 0.0395 \cdot \text{sign}\left(\dot{s}_r\right) \tag{6.33}$$

and passive joints

$$n_{pal} = -23.6001 \cdot \text{sign}\left(\dot{\theta}_l\right) \qquad n_{pll} = -9.3241 \cdot \text{sign}\left(-\dot{\theta}_l - \dot{\gamma}_l\right)$$

$$n_{par} = -23.9920 \cdot \text{sign}\left(\dot{\theta}_r\right) \qquad n_{plr} = -5.2453 \cdot \text{sign}\left(-\dot{\theta}_r - \dot{\gamma}_r\right)$$

$$n_{pM} = -2.6301 \cdot \text{sign}\left(\dot{\gamma}_l + \dot{\gamma}_r\right). \tag{6.34}$$

In order to identify the rigid body parameters $\boldsymbol{p}_{rb,min,dem}$, cp. Tab. 6.1, of the

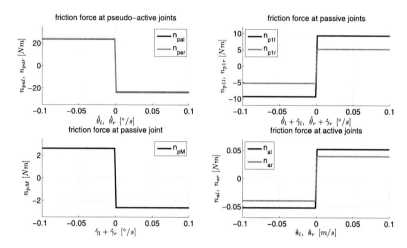

Figure 6.8: Friction models for the active, pseudo-active and passive joints

inverse dynamic model the trajectory presented in Fig. 6.2 is implemented as motion of the active joints $\boldsymbol{q}_a$ in order to optimally excite the system. The $\boldsymbol{\tau}$'s and $\boldsymbol{q}_a$'s are measured in order to determine $\boldsymbol{\psi}_{rb}$, cp. Eq. (6.24). The measurement vector $\boldsymbol{\Gamma}_{rb}$, cp. Eq. (6.24), is calculated using $\boldsymbol{\tau}$ and subtracting the afore identified friction influence. After this the rigid body parameters $\boldsymbol{p}_{rb,min}$ are identified using the LS-estimator (6.25).

The verification trajectory is implemented for testing the proposed identification method. The resulting forces $\boldsymbol{\tau}$ are compared to the inverse dynamic model with identified rigid body $\boldsymbol{p}_{rb,min,dem}$ and joint friction parameters $\boldsymbol{p}_{f,min,dem}$. The results are given in Fig. 6.9.

The differences between the model and the measurements are minimal. The simplified inverse dynamic model with identified parameters almost perfectly represents the dynamic behavior of the manipulator inside the workspace.

6.5.5 Physical interpretability

In the context of parameter identification the physical interpretability or feasibility of the determined parameters is a commonly discussed point. As above also in other publications, e.g. [Gro03, GHA04a, DRMFP08] the identification procedure sometimes leads to parameters with unfeasible values – which are in some cases

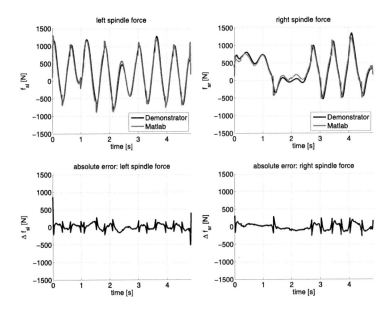

Figure 6.9: Inverse dynamic with estimated parameters compared to manipulator's behavior for verification trajectory

also eliminated [GHA04a]. The occurrence of physically not interpretable values of parameters has nothing to do with the quality of the identification process. Normally they are a consequence of model simplifications.

Physical feasibility means, that the parameters are consistent with their physical meaning described e.g. by Eq. (6.31). One possible way for verifying the feasibility of a base parameter set is proposed in [YK00]. Due to the fact that the kinetic energy of a system has always to be positive the following equation must be fulfilled, while recalculating the physical parameters, in order to guarantee their feasibility

$$m_i > 0 \qquad \text{and} \qquad {}^iI_{G_i} > 0. \qquad (6.35)$$

The feasibility can also directly be considered in the identification process. One possible strategy in case of parallel robots can be found in [FMAPV08].

But the physical feasibility can also be interpreted in another sense. Analyzing a minimal parameter vector its elements should allow a direct interpretation with respect to the (simplified) structure used for the derivation of its inverse dynamic model. This means, that the individual parameters can in some cases – and also covertly for the here derived vector (6.31) – be interpreted in form of (sums of):

- masses: m_i
- first moments of masses: $r_i \cdot m_i$ (r_i: radius of rotation)
- second moments of masses: $I = r_i^2 \cdot m_i$ (including Steiner components)

For these cases it should then also be possible, to identify concrete individual parts or combinations of parts of the (simplified) structure from the vector elements.

In general it can be said, that the better the base (parameter-linear) model used for parameter identification reproduces the robot dynamics the more probable the identification leads to physically interpretable parameters (in the first discussed sense). But it has to be emphasized, that the loss of the physical interpretability of the rigid body parameters (cp. Table 6.1) after identification is not critical for the realization of model-based control concepts and the planning of optimal trajectories. Since these points are the primary goals of this thesis the loss of the physical interpretability in case of the SpiderMill plays no further role. In the context of the above mentioned aspects an adequate as well as efficient and real-time implementable form of the inverse dynamic model is much more important than the physical interpretability of its parameters, cp. eg. [ACC99].

6.6 Conclusions and future works

In this chapter a parameter identification strategy is given for the estimation of the rigid body and the friction parameters of the simplified model of the SpiderMill. The obtained simulative and experimental results further confirm the correctness of the developed simplified modeling strategy also in combination with parameter identification.

After designing model based control strategies in Chap. 10. Further improvements of the identification results are possible. Using more sophisticated controllers the tracking of the identification trajectories and therefore also the identification results itself can be improved. Furthermore, it has to be investigated, if other, the dynamics even better describing, analytical models directly lead to a full physical interpretability of the identified parameters.

7 Trajectory planning

In this chapter a new approach for trajectory planning is introduced. It can be applied to any kind of manipulator (serial, parallel, hybrid), whose dynamics can be expressed in some kind of Lagrangian form. The resulting trajectories are – with respect to the introduced constraints – path-constrained, smooth and time-optimal trajectories (PCSMTOPT).

The problem of trajectory planning can be decomposed into two tasks: the path planning and the actual trajectory planning problem. The path planning means to define a path in the workspace of the robot to move its end-effector from an initial pose to a final one, while avoiding collisions as well as singular configurations. In case of industrial applications, normally the path is predefined by the task and has not to be planned applying an individual strategy. Whereas the path defines only geometric relationships (a series of poses) no information about their time context is given. In case of trajectory planning the time history and therefore also the derivatives of the end-effector poses or its active joint coordinates, respectively, get defined along the path.

In general, the trajectory planning problem is formulated as an optimization problem considering aspects like time- or energy-optimality in the cost function. Especially for the planning of time-optimal trajectories several strategies are available and the most important ones of it are discussed below. Also in this thesis the problem of time-optimal trajectory planning is studied, since time-optimal planned trajectories define some kind of limit case for the trajectory tracking control. Due to the fact, that none of the available strategies can be directly applied for the planning of PCSMTOPTs for a parallel robot a new strategy is necessary. The strategy proposed in this chapter can be applied to move the end-effector of a (parallel) robot time-optimal, but also vibration-free (smooth) along a predefined path. In this context the term smoothness is related to the courses (switching) of the torques, which have to be continuous. The new strategy considers all essential constraints and designs the trajectories in the phase-space.

In Sec. 7.1 the problem of designing PCSMTOPTs is introduced. After a motivation based on commonly known approaches[1] a new strategy for planning PCSM-TOPTs in the phase-space is introduced in Sec. 7.2 and applied in Sec. 7.3 for the

---

[1] For a compact presentation, the basic strategies are summarized and analyzed in App. C.

trajectory planning in case of the SpiderMill. Possible extensions of the strategy are discussed in Sec. 7.4. Conclusions are drawn in Sec. 7.5.

7.1 Problem formulation

7.1.1 General problem formulation

In general the PCSMTOPT problem is also an optimization problem which has to be solved for minimizing the traveling time t_f of the TCP between its initial (0) and final (f) pose along the path. It starts with the definition of a cost function J

$$\textbf{cost function:} \quad \min_{\mathbf{u}(t) \in \mathcal{C}} J = \int_0^{t_f} 1 \; \mathrm{dt} \; \Leftrightarrow \; \min_{\mathbf{u}(t) \in \mathcal{C}} t_f, \qquad (7.1)$$

which a smooth and time-optimal trajectory is to be planned based on. The time-optimality is directly defined by the cost function, whereas all other requests, the path and the smoothness are defined by the constraints (CT) summarized in the set of constraints (CT $\in \mathcal{C}$). The actuation variable for minimizing t_f is the vector $\mathbf{u}(t)$, which is related to the drives of the robot.

In the following all necessary CTs for the planning of PCSMTOPTs are introduced. The first is the path, which has to be a continuous C^1-curve

$$\textbf{path CT:} \quad \mathcal{B} = \left\{ {}^B\mathbf{x}_{TCP} \in \mathbb{R}^3 \; \middle| \; {}^B\mathbf{x}_{TCP} = \mathbf{r}\,(s)\,, \quad s = s_0 \ldots s_f \right\}, \qquad (7.2)$$

where ${}^B\mathbf{x}_{TCP}$ is the end-effector pose with respect to the base coordinate system. The path restricts the DOF of the end-effector to one [PJ87], namely the coordinate of the path variable[2] s.

Additionally the physical limitations of the manipulator, especially that of the drives, have to be considered. The drives in the active joints $\dot{q}_a$ have a lower and an upper speed limit

$$\textbf{velocity CT:} \quad \dot{\mathbf{q}}_{a,min} \leq \dot{\mathbf{q}}_a(t) \leq \dot{\mathbf{q}}_{a,max}. \qquad (7.3)$$

In the starting- and end-point the velocities of the active joints should be zero. The same holds for the accelerations, since for smooth trajectories a soft run-up and braking is required.

$$\textbf{start-point* CT}^3\textbf{:} \quad \dot{\mathbf{q}}_a\,(t=0) = \mathbf{0} \quad \text{and} \quad \ddot{\mathbf{q}}_a\,(t=0) = \mathbf{0}, \qquad (7.4\text{a})$$

$$\textbf{end-point* CT:} \quad \dot{\mathbf{q}}_a\,(t=t_f) = \mathbf{0} \quad \text{and} \quad \ddot{\mathbf{q}}_a\,(t=t_f) = \mathbf{0}. \qquad (7.4\text{b})$$

---

[2]The choice of s as path variable is standard in the trajectory planning literature. Due to its restriction to this chapter no confusion with the spindle length s_k occurs.

Furthermore, the torques of the drives $\boldsymbol{\tau}$ have to be within a defined interval

$$\text{torque* CT:} \quad \boldsymbol{\tau}_{min} \leq \boldsymbol{\tau}(t) \leq \boldsymbol{\tau}_{max}. \tag{7.5}$$

The limitations discussed so far have to be considered for the planning of time-optimal – and 'a little bit more smooth', than bang-bang – trajectories in the run-up and braking phase. But in order to plan smooth trajectories it has to be ensured, that the course of torque along the path is not jumping or even of bang-bang type. In order to guarantee this, the jerk[4] $\dot{\boldsymbol{\tau}}$ has also to be limited

$$\text{jerk* CT:} \quad \dot{\boldsymbol{\tau}}_{min} \leq \dot{\boldsymbol{\tau}}(t) \leq \dot{\boldsymbol{\tau}}_{max}. \tag{7.6}$$

These are all necessary constraints for the planning of (PC)SMTOPTs.

The transformation of the trajectory planning problem into an optimization problem (7.1) requires a dynamic model of the respective manipulator. Several 'general' forms of the dynamic equations (cp. Sec. 5.3) can be found in literature. Without loss of generality the dynamic model of the SpiderMill (5.25) is used in the following steps like in case of [CC00] neglecting friction terms

$$\boldsymbol{\tau} = \mathbf{M}(\mathbf{q}) \cdot \ddot{\mathbf{q}}_a + \mathbf{C}(\mathbf{q}) \cdot \dot{\mathbf{q}}^2 + \mathbf{G}(\mathbf{q}) \tag{7.7}$$

If friction will be considered the related discontinuities would lead to further problems, cp. Sec. 7.4.2. Since the jerk $\dot{\boldsymbol{\tau}}$ is neither an input nor a state, nor in form of an algebraic equation part of the dynamic model, it is not suitable for the planning of PCSMTOPTs. Therefore Eq. (7.7) is derived with respect to time [Con98]

$$\dot{\boldsymbol{\tau}} = \mathbf{M}(\mathbf{q}) \cdot \dddot{\mathbf{q}}_a + \dot{\mathbf{M}}(\mathbf{q}) \cdot \ddot{\mathbf{q}}_a + \dot{\mathbf{C}}(\mathbf{q}) \cdot \dot{\mathbf{q}}^2 + 2 \cdot \mathbf{C}(\mathbf{q}) \cdot \dot{\mathbf{q}} \cdot \ddot{\mathbf{q}} + \dot{\mathbf{G}}(\mathbf{q}). \tag{7.8}$$

Solving Eq. (7.8) for $\dddot{\mathbf{q}}_a$ yields [Con98]

$$\dddot{\mathbf{q}}_a = \mathbf{M}^{-1} \cdot \dot{\boldsymbol{\tau}} - \mathbf{M}^{-1} \cdot \underbrace{\left(\dot{\mathbf{M}} \cdot \ddot{\mathbf{q}}_a + \dot{\mathbf{C}} \cdot \dot{\mathbf{q}}^2 + 2 \cdot \mathbf{C} \cdot \dot{\mathbf{q}} \cdot \ddot{\mathbf{q}} + \dot{\mathbf{G}} \right)}_{\Pi}. \tag{7.9}$$

From which the following state-space model can be derived [Con98]:

states: $\mathbf{x}_1 = \mathbf{q}_a$, $\mathbf{x}_2 = \dot{\mathbf{q}}_a$, $\mathbf{x}_3 = \ddot{\mathbf{q}}_a$ input: $\mathbf{u} = \dot{\boldsymbol{\tau}}$ output: $\mathbf{y} = \mathbf{q}_a$

model: $\begin{bmatrix} \dot{\mathbf{x}}_1 \\ \dot{\mathbf{x}}_2 \\ \dot{\mathbf{x}}_3 \end{bmatrix} = \begin{bmatrix} \mathbf{x}_2 \\ \mathbf{x}_3 \\ -\mathbf{M}^{-1} \cdot \Pi \end{bmatrix} + \begin{bmatrix} 0 \\ 0 \\ \mathbf{M}^{-1} \end{bmatrix} \cdot \mathbf{u}, \qquad \mathbf{y} = \mathbf{x}_1.$ $\tag{7.10}$

---

[3]The $*$ indicates that the CT is not completely defined at this stage - here for example the conditions for the generalized joint variables at the start and end time are missing.

[4]If not other specified, the term *jerk* names the time derivative of the torque in this thesis.

The input of the model (7.10) and consequently the actuation variable of the optimization problem (7.1) is the jerk

$$\mathbf{u} = \dot{\boldsymbol{\tau}} : [t_0, t_f] \longrightarrow \mathcal{C} \subset \mathbb{R}^n. \tag{7.11}$$

For an appropriate course of $\dot{\boldsymbol{\tau}}$ considering the limitations t_f can be minimized and the trajectory planning problem can be solved.

7.1.2 Path dependent problem formulation

Above the general PCSMTOPT problem has been introduced giving an idea of the related problems. Beside the poses of the TCP also the kinematic and dynamic relationships and all other CTs can be expressed by the path variable s and its time-derivatives. This simplifies the solution of the problem drastically.

Parametrized kinematics and dynamics

The active joint coordinates $\mathbf{q}_a$ in dependency of s are obtained from the inverse kinematics $\mathcal{I}(\mathbf{x})$ and the path (7.2).

$$\mathbf{q}_a(t) = \mathcal{I}(\mathbf{x}) = \mathcal{I}(\mathbf{r}(s)) \tag{7.12}$$

The relationship for the velocities can be derived based on the Jacobian matrix

$$\dot{\mathbf{r}}(s) = \mathbf{J}^{-1}(\mathbf{q}_a(t)) \cdot \dot{\mathbf{q}}_a(t) \tag{7.13}$$

For non-singular configurations [PJ87, Con98]

$$\dot{\mathbf{q}}_a = \mathbf{J} \cdot \dot{\mathbf{r}} = \mathbf{J} \cdot \frac{\partial \mathbf{r}}{\partial s} \cdot \frac{\partial s}{\partial t} = \mathbf{J} \cdot \mathbf{r}' \cdot \dot{s} = \mathbf{q}_a' \cdot \dot{s} \tag{7.14a}$$

$$\ddot{\mathbf{q}}_a = \mathbf{q}_a'' \cdot \dot{s}^2 + \mathbf{q}_a' \cdot \ddot{s} \tag{7.14b}$$

$$\dddot{\mathbf{q}}_a = \mathbf{q}_a''' \cdot \dot{s}^3 + 3 \cdot \mathbf{q}_a'' \cdot \dot{s} \cdot \ddot{s} + \mathbf{q}_a' \cdot \dddot{s} \qquad \text{with } (\,\cdot\,)' = \partial/\partial s \tag{7.14c}$$

hold. The time derivatives of the path variable $\dot{s}$, $\ddot{s}$, and $\dddot{s}$ are the so-called *pseudo* values. For $\mathbf{q}_a'$, $\mathbf{q}_a''$ and $\mathbf{q}_a'''$ [PJ87, Con98]

$$\mathbf{q}_a' = \mathbf{J} \cdot \mathbf{r}' \tag{7.15a}$$

$$\mathbf{q}_a'' = \mathbf{J} \cdot \left(\mathbf{r}'' - \frac{\partial \mathbf{J}^{-1}}{\partial s} \cdot \mathbf{q}_a' \right) \tag{7.15b}$$

$$\mathbf{q}_a''' = \mathbf{J} \cdot \left(\mathbf{r}''' - \frac{\partial^2 \mathbf{J}^{-1}}{\partial s^2} \cdot \mathbf{q}_a' - 2 \cdot \frac{\partial \mathbf{J}^{-1}}{\partial s} \cdot \mathbf{q}_a'' \right) \tag{7.15c}$$

are valid. Analog relationships can also be derived for the passive coordinates. Based on the relationships above the kinematics and dynamics of a manipulator can be expressed in terms of the path variable and its time-derivatives – completely describing the movement of the manipulator along the path [Con98]

$$\boldsymbol{\tau} = \mathbf{m}\left(s\right) \cdot \ddot{s} + \mathbf{c}\left(s\right) \cdot \dot{s}^2 + \mathbf{g}\left(s\right) \tag{7.16}$$

$$\dot{\boldsymbol{\tau}} = \overline{\mathbf{m}}\left(s\right) \cdot \dddot{s} + \overline{\mathbf{c}}\left(s, \dot{s}\right) \cdot \ddot{s} + \overline{\mathbf{g}}\left(s, \dot{s}\right) \tag{7.17}$$

with

$$\mathbf{m}\left(s\right) = \mathbf{M} \cdot \mathbf{q}_a'$$
$$\mathbf{c}\left(s\right) = \mathbf{M} \cdot \mathbf{q}_a'' + \mathbf{C} \cdot \mathbf{q}'^2$$
$$\mathbf{g}\left(s\right) = \mathbf{G}$$

and

$$\overline{\mathbf{m}}\left(s\right) = \mathbf{M} \cdot \mathbf{q}_a'$$
$$\overline{\mathbf{c}}\left(s\right) = \left(3\,\mathbf{M} \cdot \mathbf{q}_a'' + \frac{\partial \mathbf{M}}{\partial s} \cdot \mathbf{q}_a' + 2\,\mathbf{C} \cdot \mathbf{q}'^2\right) \cdot \dot{s}$$
$$\overline{\mathbf{g}}\left(s\right) = \left(\mathbf{M} \cdot \mathbf{q}_a''' + \frac{\partial \mathbf{M}}{\partial s}\mathbf{q}_a'' + \frac{\partial \mathbf{C}}{\partial s} \cdot \mathbf{q}'^2 + 2\,\mathbf{C} \cdot \mathbf{q}' \cdot \mathbf{q}''\right) \cdot \dot{s}^3 + \frac{\partial \mathbf{G}}{\partial s} \cdot \dot{s}.$$

Based on the jerk equation (7.17) the dynamics of the manipulator can be expressed in the state space of s. Thereto Eq. (7.17) is solved for $\dddot{s}$ yielding [Con98]

$$\dddot{s} = \overline{\mathbf{m}}^{-1}\left(s\right) \cdot \left(\dot{\boldsymbol{\tau}} - \overline{\mathbf{c}}\left(s, \dot{s}\right) \cdot \ddot{s} - \overline{\mathbf{g}}\left(s, \dot{s}\right)\right). \tag{7.18}$$

As new states the path variable $s = x_1$, the pseudo-velocity $\dot{s} = x_2$ and the pseudo-acceleration $\ddot{s} = x_3$ and as new input the pseudo-jerk $\dddot{s} = u$ are defined, yielding the following state-space model of the manipulator [Con98]

$$\begin{bmatrix} \dot{x}_1 \\ \dot{x}_2 \\ \dot{x}_3 \end{bmatrix} = \begin{bmatrix} x_2 \\ x_3 \\ 0 \end{bmatrix} + \begin{bmatrix} 0 \\ 0 \\ 1 \end{bmatrix} \cdot u, \tag{7.19}$$

completely describing its dynamics along the path.

Path dependent optimization problem

Based on the 'new' kinematics and dynamics the PCSMTOPT problem changes

$$\text{cost function:} \quad \min_{u(t) \in \boldsymbol{\mathcal{C}}} \ t_f$$

$$\text{path CT:} \quad \mathcal{B} = \left\{ {}^{B}\mathbf{x}_{TCP,BI} \in \mathbb{R}^3 \mid {}^{B}\mathbf{x}_{TCP,BI} = \mathbf{r}(s), \quad s = s_0 \ldots s_f \right\}$$

$$\text{velocity CT:} \quad \dot{\mathbf{q}}_{a,min} \leq \dot{\mathbf{q}}_a \leq \dot{\mathbf{q}}_{a,max}, \quad \dot{\mathbf{q}}_a = \mathbf{q}_a' \cdot \dot{s}$$

$$\text{torque CT:} \quad \boldsymbol{\tau}_{min} \leq \boldsymbol{\tau} \leq \boldsymbol{\tau}_{max}, \quad \boldsymbol{\tau} = \mathbf{m}(s) \cdot \ddot{s} + \mathbf{c}(s) \cdot \dot{s}^2 + \mathbf{g}(s)$$

$$\text{jerk CT:} \quad \dot{\boldsymbol{\tau}}_{min} \leq \dot{\boldsymbol{\tau}} \leq \dot{\boldsymbol{\tau}}_{max}, \quad \dot{\boldsymbol{\tau}} = \overline{\mathbf{m}}(s) \cdot \dddot{s} + \overline{\mathbf{c}}(s,\dot{s}) \cdot \ddot{s} + \overline{\mathbf{g}}(s,\dot{s})$$

$$\text{start-point CT:} \quad \mathbf{q}_a(s_0) = \mathbf{q}_{a,0} \quad \dot{\mathbf{q}}_a(s_0) = \mathbf{0} \quad \ddot{\mathbf{q}}_a(s_0) = \mathbf{0}$$

$$\text{end-point CT:} \quad \mathbf{q}_a(s_f) = \mathbf{q}_{a,f} \quad \dot{\mathbf{q}}_a(s_f) = \mathbf{0} \quad \ddot{\mathbf{q}}_a(s_f) = \mathbf{0}. \tag{7.20}$$

For solving it the model (7.19) is applied. Based on the choice of u as actuation variable a smooth, time-optimal trajectory along a predefined path can be planned

$$u = \dddot{s} : [t_0, t_f] \longrightarrow \boldsymbol{\mathcal{C}} \subset \mathbb{R}. \tag{7.21}$$

Compared with Eq. (7.11) the actuation variable is now just a scalar, leading to a significant simplification of the optimization problem. Furthermore the transformation of the model from the state space of the joint variables to that of s allows an easier design of PCSMTOPTs.

So far the above performed steps are an extension of the approach in [Con98,CC00] by a further CT on the velocities and the extension to parallel robots with its pseudo-active and passive joint variables. Also in the following section commonly known relationships and results – recapitulated in almost all papers and books concerning the trajectory planning in the phase space – are introduced and where ever required extended. This theoretical basis is necessary in order to be able to introduce the new approach for the planning of PCSMTOPTs in Sec. 7.2.

7.1.3 Admissible region (AR)

The admissible region (AR) is the area in which all trajectories, which can be planned for the path (7.2), must lie. It is defined by the limitations of the velocity, the torque and the jerk which have to be analyzed separately. The formulation of the problem (7.20) in dependency of s allows an easier design of it.

The velocity limitation (V)

The velocity CT of Eq. (7.20) defines a lower and an upper bound for $\dot{\mathbf{q}}_a$ and therefore for the speed of the drives. Thereby $\dot{s}$ at each path point s is limited. Since a time-optimal trajectory is to be planned only its upper bound is of interest

$$\dot{s} \leq \dot{s}_{max,V}(s). \tag{7.22}$$

The index V indicates, that this limitation of $\dot{s}$ is a consequence of the velocity CTs of the active joints. For determining the upper limit of $\dot{s}$ the velocity CT of Eq. (7.20) is analyzed for each active joint

$$\dot{\mathbf{q}}_{a,i,min} \leq \mathbf{q}'_a \cdot \dot{s} \leq \dot{\mathbf{q}}_{a,i,max} \quad \text{with} \quad i = 1 \ldots n. \tag{7.23}$$

In each path point the sign of $\mathbf{q}'_a$ defines, if the upper limit $\dot{s}_{i,max}$ of the ith joint is determined by the maximal or the minimal joint velocity. The minimums of the upper bounds $\dot{s}_{i,max,V}$ define the course of the maximal pseudo-velocity $\dot{s}_{max,V}$

$$\dot{s}_{max,V}(s) = \min_i \left(\max_{\mathbf{r}(s)} \left(\frac{\dot{\mathbf{q}}_{a,i}}{\mathbf{q}'_a} \right) \right) \quad \text{with} \quad i = 1 \ldots n. \tag{7.24}$$

This curve in the s-$\dot{s}$ phase plane is called V-velocity-limitation, since it is defined by the limits of the drive velocities of the active joints. In Fig. 7.1 its design is sketched. Physically this means, that in each path point s at least one of the drives

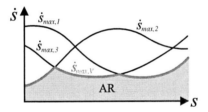

Figure 7.1: V-veloctiy-limitation

of the manipulator has to be on its upper or lower speed limit.

The torque limitation (T)

The torque CT of Eq. (7.20) defines a lower and an upper bound for the torque of the drives of the active joints. For the ith drive the following inequality arise [PJ87]

$$\boldsymbol{\tau}_{i,min} \leq \mathbf{m}_i(s) \cdot \ddot{s} + \mathbf{c}_i(s) \cdot \dot{s}^2 + \mathbf{g}_i(s) \leq \boldsymbol{\tau}_{i,max} \quad \text{with} \quad i = 1 \ldots n. \tag{7.25}$$

The inequality of the ith joint indicates, that for a constant s and for $\dot{s} \geq 0$ a couple of lines (only for the frictionless case) is defined in the $\dot{s}^2$-$\ddot{s}$ phase plane, by the two torque limits. If the couples of lines for all joints become plotted in the phase plane, a diagram like that in Fig. 7.2 (left) is obtained for a fixed s. The figure communicates an expression of the available acceleration and deceleration capacities of the robot along the path. The highlighted AR defines for each $\dot{s}$ the

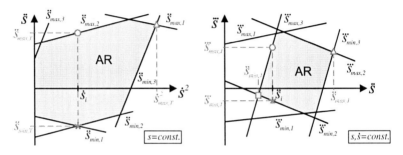

Figure 7.2: Admissible regions in the phase planes [PJ87, CC00]

interval of allowed $\ddot{s}$ [SL92]

$$\ddot{s}_{min,T}\left(s, \dot{s}\right) \leq \ddot{s} \leq \ddot{s}_{max,T}\left(s, \dot{s}\right).\qquad(7.26)$$

The index T indicates, the limit as a consequence of the torque limits. The area in which $\ddot{s}$ is defined is called T-acceleration-area. Its edges are given by the line couples. The upper limit for $\ddot{s}$ in the point $\dot{s}$ is determined by the lowest point on the upper limit straights (o) and consequently its lower limit by the largest value ($*$) on the lower limit straights [SL92].

$$\ddot{s}_{min,T}\left(s, \dot{s}\right) = \max_i\left(\min_{\tau_i}\left(\frac{\tau_i - \mathbf{c}_i\left(s\right) \cdot \dot{s}^2 - \mathbf{g}_i\left(s\right)}{\mathbf{m}_i\left(s\right)}\right)\right),\quad i = 1\ldots n\qquad(7.27a)$$

$$\ddot{s}_{max,T}\left(s, \dot{s}\right) = \min_i\left(\max_{\tau_i}\left(\frac{\tau_i - \mathbf{c}_i\left(s\right) \cdot \dot{s}^2 - \mathbf{g}_i\left(s\right)}{\mathbf{m}_i\left(s\right)}\right)\right),\quad i = 1\ldots n\qquad(7.27b)$$

with $\mathbf{m}_i \neq 0$. If $\mathbf{m}_i\left(s\right) = 0$ holds, the ith actuator does not contribute to the limit of $\ddot{s}$ in point s, the acceleration bounds are then determined by the other $n-1$ actuators. The case $\mathbf{m}_i = 0$ for all i is improbable since $\mathbf{m}$ is in general positive definite, except for some special cases discussed in [Agr91, Agr93].[5] From Fig. 7.2

---

[5]There it has been proven, that a singular configuration of the inertia matrix $\mathbf{m}$ can only occur when point masses are used for the modeling of the links. The singularities are a pure result of

(left) it can be seen, that in each point s a limit for $\dot{s}$ is defined by the AR [SL92]

$$\dot{s} \leq \dot{s}_{max,T}(s). \tag{7.28}$$

The maximum $\triangle$ is defined by the point of intersection of two certain couples of lines in the $\dot{s}^2$-$\ddot{s}$ phase plane [SL92]

$$\dot{s}^2_{max,T}(s) = \min_{i,j} \left(\max_{\tau_i, \tau_j} \left(\frac{\mathbf{m}_j(s) \cdot (\boldsymbol{\tau}_i - \mathbf{g}_i(s)) - \mathbf{m}_i(s) \cdot (\boldsymbol{\tau}_j - \mathbf{g}_j(s))}{\mathbf{c}_i(s) \cdot \mathbf{m}_j(s) - \mathbf{c}_j(s) \cdot \mathbf{m}_i(s)} \right) \right) \tag{7.29}$$

with $\mathbf{c}_i \cdot \mathbf{m}_j - \mathbf{c}_j \cdot \mathbf{m}_i \neq 0$ and $i, j = 1 \ldots n$. Otherwise the ith and jth actuator do not contribute to the limitation of $\dot{s}$ in point s or no such limit exists (infinity). An example for it is a Cartesian system, which is moving along straight line paths [SL92]. It has no velocity limit since for that case $c_i = 0$ for all i. If the $\dot{s}_{max,T}$ for each s is plotted in the s-$\dot{s}$ phase plane the T-velocity-limit curve is obtained. In each point on this curve at least one drive is in its torque limit.

The jerk limitation (J)

In [Con98, CC00] the analysis commonly applied to the torque limitations has been extended to study the influence of the jerk limits on the trajectory planning. The thereby obtained results and corresponding equations are summarized below.
The jerk CT of Eq. (7.20) defines a lower and an upper bound for the jerk of the drives of the active joints. For the ith drive the following inequality arise

$$\dot{\boldsymbol{\tau}}_{i,min} \leq \overline{\mathbf{m}}_i(s) \cdot \dddot{s} + \overline{\mathbf{c}}_i(s, \dot{s}) \cdot \ddot{s} + \overline{\mathbf{g}}_i(s, \dot{s}) \leq \dot{\boldsymbol{\tau}}_{i,max}, \quad i = 1 \ldots n. \tag{7.30}$$

The inequality of the ith joint indicates, that for a constant s and a constant $\dot{s}$ a couple of lines (only for the frictionless case) is defined in the $\ddot{s}$-$\dddot{s}$ phase plane by the two limits of the jerk, cp. Fig. 7.2 (right). The AR describes two things: an interval for $\ddot{s}$ and for a concrete $\ddot{s}$ additionally an allowed range for $\dddot{s}$ [CC00].

$$\ddot{s}_{min,J}(s, \dot{s}) \leq \ddot{s} \leq \ddot{s}_{max,J}(s, \dot{s}) \tag{7.31a}$$

$$\dddot{s}_{min,J}(s, \dot{s}, \ddot{s}) \leq \dddot{s} \leq \dddot{s}_{min,J}(s, \dot{s}, \ddot{s}). \tag{7.31b}$$

The index J indicates, that the limits are a consequence of the limitation of the jerk of the drives of the active joints. The area in which $\ddot{s}$ is defined is called J-acceleration-area. The limits (min: $\square$, max: $\triangle$) of $\dddot{s}$ are defined by the straights.

---

the link geometry and the locations of the centers of masses $\mathcal{G}_i$. For example, $\mathbf{m}$ gets singular, when for a planar robot with only revolute joints the $\mathcal{G}_i$ of at least one link is located at a joint. This type of singularity is not equal to the kinematic ones in Sec. 3.4.

$$\ddot{s}_{min,J}\left(s,\dot{s},\ddot{s}\right)=\max_{i}\left(\min_{\dot{\tau}_i}\left(\frac{\dot{\tau}_i-\overline{\mathbf{c}}_i\left(s,\dot{s}\right)\cdot\ddot{s}-\overline{\mathbf{g}}_i\left(s,\dot{s}\right)}{\overline{\mathbf{m}}_i\left(s\right)}\right)\right)\qquad(7.32\text{a})$$

$$\ddot{s}_{max,J}\left(s,\dot{s},\ddot{s}\right)=\min_{i}\left(\max_{\dot{\tau}_i}\left(\frac{\dot{\tau}_i-\overline{\mathbf{c}}_i\left(s,\dot{s}\right)\cdot\ddot{s}-\overline{\mathbf{g}}_i\left(s,\dot{s}\right)}{\overline{\mathbf{m}}_i\left(s\right)}\right)\right)\qquad(7.32\text{b})$$

for $i=1\dots n$, with $\overline{\mathbf{m}}_i\neq 0$. Otherwise $\dddot{s}$ in point s is not limited by the ith actuator, but by the remaining $n-1$ ones. The two limits of $\ddot{s}$ are defined by the following straight intersection points in the $\ddot{s}$-$\dddot{s}$ phase plane

$$\ddot{s}_{min,J}\left(s,\dot{s}\right)=\max_{i,j}\left(\min_{\dot{\tau}_i,\dot{\tau}_j}\left(\frac{\overline{\mathbf{m}}_j\cdot\left(\dot{\tau}_i-\overline{\mathbf{g}}_i\right)-\overline{\mathbf{m}}_i\cdot\left(\dot{\tau}_j-\overline{\mathbf{g}}_j\right)}{\overline{\mathbf{c}}_i\cdot\overline{\mathbf{m}}_j-\overline{\mathbf{c}}_j\cdot\overline{\mathbf{m}}_i}\right)\right)\qquad(7.33\text{a})$$

$$\ddot{s}_{max,J}\left(s,\dot{s}\right)=\min_{i,j}\left(\max_{\dot{\tau}_i,\dot{\tau}_j}\left(\frac{\overline{\mathbf{m}}_j\cdot\left(\dot{\tau}_i-\overline{\mathbf{g}}_i\right)-\overline{\mathbf{m}}_i\cdot\left(\dot{\tau}_j-\overline{\mathbf{g}}_j\right)}{\overline{\mathbf{c}}_i\cdot\overline{\mathbf{m}}_j-\overline{\mathbf{c}}_j\cdot\overline{\mathbf{m}}_i}\right)\right)\qquad(7.33\text{b})$$

with $\overline{\mathbf{c}}_i\cdot\overline{\mathbf{m}}_j-\overline{\mathbf{c}}_j\cdot\overline{\mathbf{m}}_i\neq 0$ and $i,j=1\dots n$. Otherwise the ith and jth actuator do not contribute for a fixed s and $\dot{s}$ to the limitation of $\ddot{s}$ or no such limit exists.
The two limit points of $\ddot{s}$ in the $\ddot{s}$-$\dddot{s}$ phase plane mark two limit planes in the s-$\dot{s}$-$\ddot{s}$ phase space, which intersects. The projection of the curve of intersection in the s-$\dot{s}$ phase plane is called J-velocity-limit curve. It defines an upper bound for $\dot{s}$ caused by the limited jerk in the s-$\dot{s}$ phase plane [CC00]

$$\dot{s}\leq\dot{s}_{max,J}\left(s\right).\qquad(7.34)$$

This limitation of $\dot{s}$ for a given s, is the situation [CC00], where the AR in the $\ddot{s}$-$\dddot{s}$ plane reduces to a point $\ddot{s}_{min,J}=\ddot{s}_{max,J}$ in Fig. 7.2.

Overall admissible region (OAR)

The CTs of the pseudo values $\dot{s}$, $\ddot{s}$ and $\dddot{s}$ are limiting the AR in which all trajectories are lying, which can be planned for the path of the end-effector. Above the CTs have been individually and separately analyzed in the space of the path variable. As a result several areas of validity for $\dot{s}$, $\ddot{s}$ and $\dddot{s}$ have been obtained. In the following an overall area of validity for all three of them will be defined and based on it an OAR derived.
As a consequence of the physical limits of $\dot{\mathbf{q}}_a$, $\boldsymbol{\tau}$ and $\dot{\boldsymbol{\tau}}$ three limit curves for the $\dot{s}$ have been obtained in the s-$\dot{s}$ phase plane: the V-, the T- and the J-velocity-limit curve. From them an overall limit, the so called maximal-velocity-limit-curve (MVC) is obtained, when for a fixed s the minimal value of all three limit curves

is determined

$$MVC = \left\{ (s, \dot{s}) \subset \mathbb{R}^2 \mid s_0 \leq s \leq s_f \ \wedge \ \dot{s}(s) = \dot{s}_{max}(s) \right\}$$
$$\dot{s}_{max}(s) = \min \left(\dot{s}_{max,V}(s), \dot{s}_{max,T}(s), \dot{s}_{max,J}(s) \right). \tag{7.35}$$

From the limits of $\boldsymbol{\tau}$ and $\dot{\boldsymbol{\tau}}$ the limits for $\ddot{s}$ get derived. These are the T-acceleration-area and the J-acceleration-area. Based on it the limits of the maximal acceleration interval can be determined. For a fixed s and $\dot{s}$ its lower limit is defined by the maximum of the lower limits of the T- and J-acceleration-area. The upper limits are determined in an analogous manner [CC00]

$$\ddot{s}_{min}(s, \dot{s}) = \max \left(\ddot{s}_{min,T}(s, \dot{s}), \ddot{s}_{min,J}(s, \dot{s}) \right) \tag{7.36a}$$
$$\ddot{s}_{max}(s, \dot{s}) = \min \left(\ddot{s}_{max,T}(s, \dot{s}), \ddot{s}_{max,J}(s, \dot{s}) \right). \tag{7.36b}$$

The limits of $\dddot{s}$ are only a result of the limitation of $\dot{\boldsymbol{\tau}}$, cp. Eq. (7.32) [CC00]

$$\dddot{s}_{min/max}(s, \dot{s}, \ddot{s}) = \dddot{s}_{min/max,J}(s, \dot{s}, \ddot{s}). \tag{7.37}$$

The above introduced limits define the OAR, which is defined by

$$OAR = \{ (s, \dot{s}, \ddot{s}, \dddot{s}) \subset \mathbb{R}^4 \mid s_0 \leq s \leq s_f \ \wedge 0 \leq \dot{s} \leq \dot{s}_{max}(s) \ \wedge$$
$$\ddot{s}_{min}(s, \dot{s}) \leq \ddot{s} \leq \ddot{s}_{max}(s, \dot{s}) \ \wedge \dddot{s}_{min}(s, \dot{s}, \ddot{s}) \leq \dddot{s} \leq \dddot{s}_{max}(s, \dot{s}, \ddot{s}) \}. \tag{7.38}$$

This region is a result of the transformation of the physical limits in the space s. It defines the solution space, in which all trajectories lie, which can be planned for the path (7.2). This solution space defines the basis for the planning of PCSMTOPTs.

7.2 Design of PCSMTOPTs in phase space

7.2.1 Motivation for a new strategy

In this section it is shortly discussed, whether commonly known concepts of trajectory planning – summarized in App. C.1 – can be applied for solving the optimization problem (7.20) or if it is necessary to develop a new strategy.

Most of the approaches directly based on the algorithm of Pfeiffer and Johanni [PJ86, PJ87] only consider path and torque CTs. But the velocity and jerk CTs are not taken into account, whereas the latter one leads to a 'bang-bang' behavior of the torques and consequently not to smooth trajectories. Along the path (7.2), for every s at least one actuator is in its torque limit resulting in time-optimal trajectories. Even though the trajectories are designed in the s-$\dot{s}$ phase plane a time-domain model is used for calculating s, $\dot{s}$ and $\ddot{s}$ with respect to time. But a model defining $\dot{s}$ and $\ddot{s}$ with respect to s would be more reasonable. For example, this would simplify the calculation of a trajectory until an intersection point with a second one in the s-$\dot{s}$ phase plane.

An extension of the algorithm in the 'right' direction is that by Constantinescu and Croft considering additionally the jerk CTs, but still neglecting that of the velocities. In this concept the non-differentiable and critical points[6] of the basic approach are considered as interval boundaries, between which cubic splines are used to approximate the trajectory. The approach leads to (with respect to the approximations) PCSMTOPTs. But due to the approximations only a suboptimal solution is obtained. For solving the optimization problem a numerical optimizer is used. Normally, these optimizers only reach the global optimum, when the initial conditions have been chosen correctly. This is extremely difficult in cases with numerous optimization parameters, like for parallel robots with its large number of critical points. Here a design of the solution based on the strategy proposed from Pfeiffer and Johanni is more reasonable. Therefore, the strategy of Constantinescu and Croft is not appropriate for parallel robots.

Since all approaches do not consider all CTs of the optimization problem (7.20) – more or less simplifying it – or are not applicable to parallel robots a new strategy, extending the existing ones is necessary.

7.2.2 Fundamentals of the strategy

The basic idea of the new strategy is related to the approach of Lambrechts [Lam03, LBS05]. He described an algorithm for higher-order trajectory planning of a linear transport system (or an individual axis of a manipulator).

Higher-order trajectory planning: The strategy of Lambrechts can be applied for point-to-point (ptp) movements. Beside start- and end-point CTs further limits for the time-derivatives of the position x are taken into account. Depending on the order of the planned trajectory limits for its velocity v, acceleration a and jerk j (in case of third-order trajectory planning, cp. Fig. 7.3)[7] or additional limits for further time-derivatives of x are considered. But for the latter ones it is difficult to define physically reasonable and interpretable bounds. Based on the limitations switching instants (t_1 to t_7 in Fig. 7.3) can be calculated. These points define the intervals – when and how long a limit is active – in order to get a minimal t_f. The

---

[6]The special terminologies of the approaches are explained in App. C.1.

[7]The courses of the jerk, velocity and acceleration are normalized (N) to their maximum value. Furthermore in this context the jerk is the time-derivative of the acceleration.

calculation of the t_is can be found in [Lam03, LBS05].

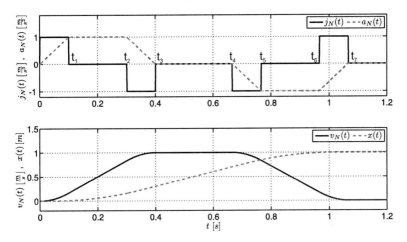

Figure 7.3: Third-order trajectory planning according to Lambrechts [Lam03]

Weaknesses of the concept in case of parallel robots: As already mentioned above, the strategy can only be applied to a single axis of a manipulator in case of ptp-motions. Consequently, after a structural decomposition of the limbs of a parallel robot the strategy can also be applied to it in principle. An example for its application to a tripod structure can be found in [KGLS07a]. But due to the existing coupling of the limbs and the low model accuracy of single axis models (cp. also Sec. 5.2) the basic strategy leads to only suboptimal results in case of parallel kinematic structures.

For the planning of PCSMTOPTs for parallel robots it is not enough to plan smooth, time-optimal ptp-motions for each active joint. Consequently the strategy of Lambrechts cannot be directly applied. Furthermore, it is also difficult to get the required information about the limitations of the accelerations of the drives or their jerks. In general they have to be determined by measurements.

Basic ideas for the new strategy: But the approach of Lambrechts provides some ideas for the planning of PCSMTOPTs in case of parallel robots. Analyzing Fig. 7.3 in more detail, it is obvious, that in case of a time-optimal trajectory at each time instant t between the starting- and end-point one of the time-derivatives of x is on its <u>lower</u> or $\overline{\text{upper}}$ limit, which is a criterion for a time-optimal movement of a single axis.

Additionally a statement concerning the planning of smooth trajectories can be made. In case of third-order trajectory planning (see above) for the movement of a single axis the term 'smooth' means that the course of the acceleration $a\,(t)$ is continuous. The smooth course is obtained, since directly before and behind an interval in which a is on its limit an interval exists in which j is in its limit.

Transfer of the basic ideas to (parallel) robots: The ideas above are used in the following to design PCSMTOPTs in the phase-space. For that purpose the kinetic quantities in case of a robot have to be determined: s is equivalent to the position x of the single-axis movement since it defines the pose of the end-effector according to Eq. (7.2). Consequently its time-derivatives $\dot{s}$, $\ddot{s}$ and $\dddot{s}$ can be regarded as equivalents for v, a and j.

If at each time-instant along the path at least one kinetic quantity is in its limitation, analogous to the approach of Lambrechts, the movement has a time-optimal characteristics – cp. also [SM85]. But the kinetic quantities of the end-effector of a robot are more difficult to handle than that of a linear axis. The required limitations can be traced back to the physical limits of the drives, as obvious from the optimization problem (7.20) and explained in detail in Sec. 7.1.3. Normally, for the actuators of the manipulator limits for the velocity $\dot{\mathbf{q}}_a$ and the torques $\boldsymbol{\tau}$ are known and the limitations of the jerk $\dot{\boldsymbol{\tau}}$ can be determined. Additionally, since a corresponds to $\ddot{s}$ also a smoothness criterion can be defined in the phase space.

Proposition 1 *For the planning of PCSMTOPTs the following two conditions have to be satisfied:*

(1) *The movement of the end-effector of a manipulator along a defined path* (7.2) *has a **time-optimal** characteristics, if at each time-instant at least one of the limits out of the set of CTs on the kinetic quantities of the n active joints $\mathcal{C}_{kin}$ is active. ($\mapsto$ PCTOPT: $\underline{p}$ath-$\underline{c}$onstrained, $\underline{t}$ime-$\underline{o}$ptimal $\underline{t}$rajectory)*

$$\mathcal{C}_{kin} = \left\{ \underline{\dot{\mathbf{q}}}_{a,1} \quad \cdots \quad \overline{\dot{\mathbf{q}}}_{a,n} \quad \underline{\tau}_{a,1} \quad \cdots \quad \overline{\tau}_{a,n} \quad \underline{\dot{\tau}}_{a,1} \quad \cdots \quad \overline{\dot{\tau}}_{a,n} \right\}. \qquad (7.39)$$

(2) *The movement of the end-effector is additionally **smooth**, if the course of the pseudo-acceleration $\ddot{s}\,(t)$ is continuous. ($\mapsto$ PCSMTOPT)*

In order to fulfill the second condition of Prop. 1 it is necessary, that the pseudo-jerk $\dddot{s}$ is limited along the whole trajectory – except in the starting- and end-point [Con98]. If the pseudo-acceleration $\ddot{s}$ is continuous along the path, this does also hold for pseudo-velocity $\dot{s}$. This has been shown in [Con98] – in reverse direction – and is recapitulated below:

If $\dot{s}$ is a feasible since continuous curve in the s-$\dot{s}$ phase plane its first derivative is

$$\ddot{s} = \frac{\partial \dot{s}}{\partial t} = \frac{\partial \dot{s}}{\partial s} \cdot \dot{s} \quad \Leftrightarrow \quad \frac{\partial \dot{s}}{\partial s} = \frac{\ddot{s}}{\dot{s}}. \tag{7.40}$$

Since both $\dot{s}$ and $\ddot{s}$ are continuous functions the continuity of $\frac{\partial \dot{s}}{\partial s}$ along the entire trajectory is given. Therefore the second condition of Prop. 1 is fulfilled, if $\dot{s}(s)$ is continuous and differentiable in the s-$\dot{s}$ phase plane [Con98].

7.2.3 Models for the trajectory planning

For the design of PCSMTOPTs in the s-$\dot{s}$ and $\dot{s}$-$\ddot{s}$ phase planes models are necessary, which transform the physical limits of the drives into the space of s. A general analysis has been performed in Sec. 7.1.3, but in case of the SpiderMill some simplifications are possible, which will be explained in the following. A further general easing results from the fact, that it is possible to integrate the following models with respect to s (instead of the time).

For solving the PCSMTOPT problem (7.20) in Sec. 7.1.3 the OAR has been studied. In following it will be restricted, by assuming that the MVC (7.35) is equal to the V-velocity-limit-curve

$$\dot{s}_{max}(s) = \dot{s}_{max,V}(s). \tag{7.41}$$

Furthermore it is assumed, that the maximal region of the pseudo-acceleration (7.36) is equivalent to the T-acceleration-area

$$\ddot{s}_{min/max}(s, \dot{s}) = \ddot{s}_{min/max,T}(s, \dot{s}). \tag{7.42}$$

For the limits of the maximal region of the pseudo-jerk no simplifications are made

$$\dddot{s}_{min/max}(s, \dot{s}, \ddot{s}) = \dddot{s}_{min/max,J}(s, \dot{s}, \ddot{s}). \tag{7.43}$$

The above introduced simplification are valid for the SpiderMill in case of the trajectories designed in Sec. 7.3.1. Analyzing them it becomes obvious, that none of it violates the assumptions. Due to the assumptions the solutions space (7.38) becomes considerably simplified

$$\mathcal{OAR}_{SM} = \{(s, \dot{s}, \ddot{s}, \dddot{s}) \subset \mathbb{R}^4 \mid s_0 \le s \le s_f \ \wedge \ 0 \le \dot{s} \le \dot{s}_{max,V}(s) \ \wedge$$

$$\ddot{s}_{min,T}(s, \dot{s}) \le \ddot{s} \le \ddot{s}_{max,T}(s, \dot{s}) \ \wedge \ \dddot{s}_{min,J}(s, \dot{s}, \ddot{s}) \le \dddot{s} \le \dddot{s}_{max,J}(s, \dot{s}, \ddot{s})\}. \tag{7.44}$$

Since the MVC is equivalent to the V-velocity-limit-curve, it is completely defined by the limits of the drive velocities $\dot{\mathbf{q}}_a$ according to Eq. (7.24). At each point of the MVC at least one drive is in its limitation. In contrast to the basic strategy of Pfeiffer and Johanni, where the MVC is only a source or sink for the trajectories, here it becomes as an result of the simplifications to a reachable curve, which can become part of the trajectories in the s-$\dot{s}$ phase plane and this not only at single points or short intervals.

Since the maximal area of $\ddot{s}$ is equivalent to the T-acceleration-area, its limits are completely defined by that of the drive torques $\boldsymbol{\tau}$ according to Eq. (7.27). For planning a trajectory in the s-$\dot{s}$ phase plane, for which $\ddot{s}$ is in its limitation, a model can be developed which has the T-acceleration-area (7.27) as an 'input'. For deriving the model $\ddot{s}$ is studied in more detail [PJ87]

$$\ddot{s} = \frac{\partial \dot{s}}{\partial s} \cdot \frac{\partial s}{\partial t} = \dot{s}' \cdot \dot{s} . \tag{7.45}$$

Furthermore also the time is studied

$$\frac{\partial s}{\partial t} = \dot{s} \quad \Leftrightarrow \quad \frac{\partial t}{\partial s} \cdot \dot{s} = 1 \quad \Leftrightarrow \quad t' \cdot \dot{s} = 1 . \tag{7.46}$$

Since a model is required for planning the trajectories in the s-$\dot{s}$ phase plane, for which $\ddot{s}$ is in its limitation, the input u is chosen according to Eq. (7.27).

states: $x_1 = \dot{s}$, $x_2 = t$, input: $u = \ddot{s} \in \{\ddot{s}_{min,T}(s, x_1, \boldsymbol{\tau}), \ddot{s}_{max,T}(s, x_1, \boldsymbol{\tau})\}$

$$\text{model:} \begin{bmatrix} x_1 & 0 \\ 0 & x_1 \end{bmatrix} \cdot \begin{bmatrix} x_1' \\ x_2' \end{bmatrix} = \begin{bmatrix} u \\ 1 \end{bmatrix} . \tag{7.47}$$

The main advantage of the model (7.47) is the use of s as integration variable. With it trajectories can be descriptively designed in the s-$\dot{s}$ phase plane.

The maximal pseudo-jerk $\dddot{s}$ (7.37) is not affected by the simplification. Its limits are completely defined by the limits of the jerk of the drives $\dot{\boldsymbol{\tau}}$ (7.32). If a trajectory shall be designed in the s-$\dot{s}$ phase plane, for which $\dddot{s}$ is in its limitation, a model can be developed, which has $\dddot{s}$ as input. Thereto, in the following the $\ddot{s}$ and $\dddot{s}$ are studied in detail

$$\ddot{s} = \frac{\partial \dot{s}}{\partial s} \cdot \frac{\partial s}{\partial t} = \dot{s}' \cdot \dot{s} \qquad \text{and} \qquad \dddot{s} = \frac{\partial \ddot{s}}{\partial s} \cdot \frac{\partial s}{\partial t} = \ddot{s}' \cdot \dot{s} . \tag{7.48}$$

For the time again Eq. (7.46) is valid. Resulting in the following state-space model, with s as integration variable

states: $x_1 = \dot{s}, \ x_2 = \ddot{s}, \ x_3 = t$

input: $u = \dddot{s} \ \in \{\dddot{s}_{min,J}(s, x_1, x_2, \dot{\boldsymbol{\tau}}), \dddot{s}_{max,J}(s, x_1, x_2, \dot{\boldsymbol{\tau}})\}$

model: $\begin{bmatrix} x_1 & 0 & 0 \\ 0 & x_1 & 0 \\ 0 & 0 & x_1 \end{bmatrix} \cdot \begin{bmatrix} x_1' \\ x_2' \\ x_3' \end{bmatrix} = \begin{bmatrix} x_2 \\ u \\ 1 \end{bmatrix}$. $\hfill (7.49)$

As a result of the above performed simplifications it is possible to infer bijectively from the limitations of the pseudo terms to the limitations of the physical values of the drives of the manipulator.

7.2.4 Trajectory planning in the phase plane

In this section PCSMTOPTs will be planned in the phase space of s. The strategy bases upon the simplifications and models of Sec. 7.2.3.

The s-$\dot{s}$ phase plane is used for the design of PCSMTOPTs. In the plane the MVC (7.41) is plotted as an upper limit curve, while in the first step a time-optimal trajectory (PCTOPT) is planned. According to [PJ87]

$$\dot{s}(s) = \frac{ds}{dt} \quad \Rightarrow \quad t_f = \int_{s_0}^{s_f} \frac{1}{\dot{s}(s)} ds \qquad (7.50)$$

for that purpose the area under the function $\dot{s}(s)$ has to be maximized. Furthermore, the requirements (7.20) defines, that the end-effector velocity and with it $\dot{s}$ in the starting- and end-point has to be zero. For that reason from starting-point s_0 it is integrated in forward direction with $\ddot{s}_{max}$ using model (7.47) until the MVC is reached. Similarly, the model is used to integrate in backward direction from the end-point s_f with $\ddot{s}_{min}$ until the MVC is reached. As result a plot like that in Fig. 7.4 is obtained.

According to Prop. 1 the planned trajectory has a time-optimal characteristic. Along curve 1 and 2 $\ddot{s}$ is in its limitation. Consequently, due to the simplifications defined in Sec. 7.2.3 it can be concluded, that in each path point along both curves at least one drive is in its torque limit. Furthermore, in each point on the MVC at least one drive is in its velocity limit. As sketched in Fig. 7.4 and verified by the example in Fig. 7.10 and 7.11 the designed trajectory in the s-$\dot{s}$-phase plane is not differentiable in each point. This implies, that $\ddot{s}$ is discontinuous. Hence the trajectory is not smooth with respect to Prop. 1.

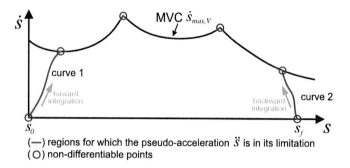

(—) regions for which the pseudo-acceleration $\ddot{s}$ is in its limitation
(O) non-differentiable points

Figure 7.4: Time-optimal trajectory in the phase plane

Non-differentiable points can occur at the changeover of two intervals, whereas in one interval $\dot{s}$ and in the other one $\ddot{s}$ is in its limitation. But also along the MVC they can occur. Although at each path point at least one drive is in its velocity limit, different drives can be in their velocity limits at the individual points, cp. Fig. 7.5. At the transition points s_T, where the drive, which is in its velocity limit,

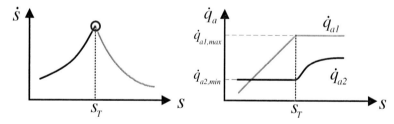

Figure 7.5: Behaviour of non-differentiable points on the MVC

is changing, the courses of the drive velocities $\dot{q}_a$ are not differentiable. Therefore the drive acceleration $\ddot{q}_a$ is not continuous as well as the belonging torque τ

$$\tau_i = 2 \cdot \pi \cdot J_i \cdot \ddot{q}_{a,i}. \qquad (7.51)$$

As a consequence, according to Eq. (7.16), $\ddot{s}$ is discontinuous in the transition points. For this reason the course of the MVC is according to Prop. 1 not smooth. Hence a way has to be found for 'smoothing' the transitions. In Fig. 7.6 this is shown graphically. Starting point is curve A and B, which intersect in the s-$\dot{s}$ phase plane. The two curves in the $\dot{s}$-$\ddot{s}$ phase plane are belonging to them. The transition

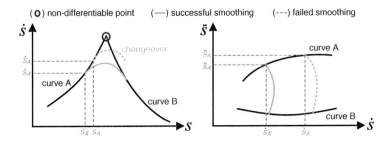

Figure 7.6: Smoothing of the MVC in the phase space

between both curves is to be smoothed. For the calculation of the transition the model in Eq. (7.49) is used. Depending on the concrete transition the minimal or the maximal pseudo-jerk is used as input. For the initialization of the model a point $(s_A, \dot{s}_A)$ (Fig. 7.6, left) on curve A near to the non-differentiable point is chosen, with its belonging $\ddot{s}_A$ (Fig. 7.6, right). Starting from $(s_A, \dot{s}_A)$ it will be integrated until the transition reaches curve B in the $\dot{s}$-$\ddot{s}$ phase plane. Afterwards it will be checked in the s-$\dot{s}$ phase plane whether the transition curve and curve B intersect in the end-point of the transition curve. A point of intersection in the $\dot{s}$-$\ddot{s}$ phase plane not necessarily leads to a point of intersection in the s-$\dot{s}$ phase plane. Although the transition curve and curve B then have the same pseudo-velocity, the transition curve can reach it at a later point on the path, cp. Fig. 7.6. If the transition curve and curve B do not intersects, then a new starting point $(s_{A'}, \dot{s}_{A'})$ on curve A has to be selected and the steps repeated, until they intersects in both phase planes. Here the model (7.49) with s as integration variable shows its strength, since the interval of the starting points for s_A as well as the breaking condition can be described much easier, than in cases where the time is used as integration variable.

Based on the results above a strategy for planning PCSMTOPTs in the phase plane of s can be introduced applying the proposed simplifications and models of Sec. 7.2.3. The procedure is shown exemplarily in Fig. 7.7.

Algorithm:

(1a) Determination of the MVC (7.41).

(1b) Smoothing of the MVC.

⇒ **step 1: MVC is a smooth reachable curve, which is traced.**

(2a) Forward-integration with maximal $\ddot{s}$ using model (7.49) until the maximal $\ddot{s}$

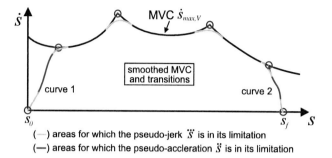

Figure 7.7: Path-constrained smooth, time-optimal trajectory (PCSMTOPT)

is reached. Afterwards, forward-integration with maximal $\ddot{s}$ using model (7.47) until the MVC gets intersected. ($\mapsto$ curve 1)

(2b) Backward-integration with minimal $\dddot{s}$ using model (7.49) until the minimal $\ddot{s}$ is reached. Afterwards, backwar-integration with minimal $\ddot{s}$ using model (7.47) until the MVC (7.41) gets intersected. ($\mapsto$ curve 2)

(2c) Smoothing of the transitions to the MVC using model (7.49).

$\Rightarrow$ **step 2: The smooth connection to/from the traced MVC is found.**

With this strategy a differentiable course is obtained in the s-$\dot{s}$ phase-plane and hence according to Prop. 1 a smooth trajectory. In each point along the trajectory one of the physical parameters responsible for the movement is in its limitation. Consequently, the trajectory has according to Prop. 1 a time-optimal characteristic. With this strategy PCSMTOPTs can be planned in the phase plane of s.

7.3 PCSMTOPTs for the SpiderMill

For comparing the control strategies in Chap. 10 in case of practice oriented situations end-effector paths of various geometry have to be planned, before the belonging trajectories can be designed. In this section the reduced Newton-Euler model of the SpiderMill, derived in Sec. 5.1.3, is applied. For simplification and ensuring a descriptive interpretability in the phase plane friction is neglected.

7.3.1 Path planning

The end-effector paths has to be planned in the workspace of the SpiderMill, especially considering the sensor-workspace limited by the laser sensors, cp. Sec. 3.3. In order to excite different parts of the system dynamics the following paths, as also proposed by [Pie03], are planned.

circle path: With a circle path the system behavior in case of a continuous movement can be tested. Along the path the TCP performs a continuous change of its direction, which enforces the influences of the centrifugal and Coriolis forces on system dynamics. Therefore circle paths are a particular challenge for robot control.

triangle path: In case of many industrial applications a manipulator has to move along straights. In order to define a closed path three straights become combined to a triangle. The movement of the end-effector along the triangle is an interrupted one. The robot is braked to zero at the end of each straight. In contrast to the circle path, no continuous change of direction occurs along a straight. For that reason the influence of non-linear dynamic effects is small in contrast to the dominating effect of the gravitation.

eight path: The lying eight, is planned as closed movement like the circle path. It is the combination of the circle elements and the straights, whereas the latter ones are combined with the circle segments in a tangential manner. For this path both types of influences play a role. Of special interest are the transitions between the two kinds of elements. In contrast to the triangle the robot is not braked to zero speed at the end of the straights or circle segments, respectively. Consequently acceleration jumps occur at these points.

All three paths are shown in Fig. 7.8 with their starting points ($\diamond$). Their running direction is counter-clockwise and they are planned with a safety distance from the boundary of the sensor workspace.

7.3.2 Trajectory planning for the SpiderMill

In this section PCSMTOPTs are planned for the paths defined in Sec. 7.3.1 applying the strategy of Sec. 7.2. This will be done exemplarily for the circle path. The particularities of the other paths will only shortly be mentioned.

In the first step the physical limits of the active joints of the SpiderMill have to be defined. They are summarized in Tab. 7.1. Since there is no exact knowledge about the jerk limits of the drives, they are taken in the range of literature values [CC00] (similar decimal powers, but values related to limits assumed for the

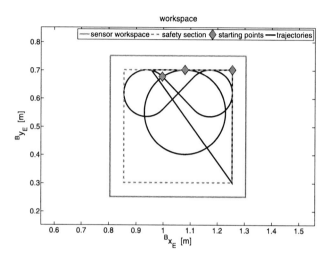

Figure 7.8: Paths of the end-effector of the SpiderMill

torques). As discussed above and in [Con98, CC00] the choice of the jerk level has a direct impact on the smoothness of the trajectories: The lower the chosen jerk limit is, the smoother the corresponding torques are. Consequently trajectories planned for a lower limit can be tracked better by a controller (cp. Sec. 10.3.4), whereas on the on the other hand the traveling time increases (cp. Tab. 7.2). Furthermore, the wear and the strain on the actuators decrease with the jerk limit.

For the trajectory planning the defined limits of the drives have been chosen consciously lower than the physical limits, in order to ensure that the drives have still sufficient control energy for the trajectory tracking control in Chap. 10.

In the following the planning of the circle trajectory for jerk limit of $|\dot{\tau}| \leq 1.6 \, \frac{\mathrm{Nm}}{\mathrm{s}}$ will be discussed exemplarily. For the design of the PCSMTOPT in the phase plane of s the models (7.47) and (7.49) are used. Their inputs are calculated in dependency of the torque τ and the jerk $\dot{\tau}$ applying Eq. (7.27) and (7.32). In order to be able to determine the courses of the inputs the vectors $\mathbf{m}(s)$, $\mathbf{c}(s)$, $\mathbf{g}(s)$ and $\overline{\mathbf{m}}(s)$, $\overline{\mathbf{c}}(s)$, $\overline{\mathbf{g}}(s)$ as well as the derivative of $\mathbf{m}'(s)$, $\mathbf{c}'(s)$, $\mathbf{g}'(s)$ with respect to s have to be determined. Whereas the first ones can be directly calculated, it is difficult to determine the latter ones analytically. Therefore they will be approximated using differential quotients:

$$y_i' = \frac{y_{i+1} - y_i}{x_{i+1} - x_i} \quad \text{(at the border)} \quad , \quad y_i' = \frac{y_{i+1} - y_{i-1}}{x_{i+1} - x_{i-1}} \quad \text{(between)}. \quad (7.52)$$

	physical limits	defined limits
rotational speed	$-3000 \frac{1}{\min} \ldots + 3000 \frac{1}{\min}$	$-1000 \frac{1}{\min} \ldots + 1000 \frac{1}{\min}$
torque	$-3.2 \, \mathrm{Nm} \ldots + 3.2 \, \mathrm{Nm}$	$-1.6 \, \mathrm{Nm} \ldots + 1.6 \, \mathrm{Nm}$
jerk	unknown	$-1.6 \frac{\mathrm{Nm}}{\mathrm{s}} \ldots + 1.6 \frac{\mathrm{Nm}}{\mathrm{s}}$ $-16 \frac{\mathrm{Nm}}{\mathrm{s}} \ldots + 16 \frac{\mathrm{Nm}}{\mathrm{s}}$ $-\infty \ldots + \infty$

Table 7.1: Physical and defined limits of the drives for trajectory planning

The vectors $\mathbf{m}(s) = \overline{\mathbf{m}}(s)$ (cp. Eq. (7.16) and (7.17)) have to be analyzed for zero values, since they are the denominators of the fractures in Eq. (7.27) and (7.32). In Fig. 7.9 their course is shown for the circle path. With the knowledge of the

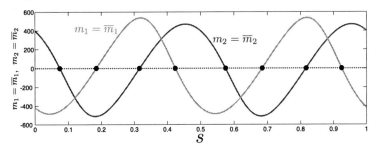

Figure 7.9: Elements of the vectors $\mathbf{m}(s) = \mathbf{m}'(s)$ along the circle path

zero values the integrations in case of the models (7.47) and (7.49) can be better initialized.

The fist step in the design of a PCSMTOPT is the calculation of the MVC. In Fig. 7.10 it is shown for the circle in the s-$\dot{s}$ phase plane. For the MVC the path acceleration $\ddot{s}$ has to be calculated. This will be done approximatively. Thereto a time base for the MVC is determined with

$$t(s) = \int_0^s \frac{1}{\dot{s}_{max}(s)} ds. \tag{7.53}$$

Applying the differential quotients (7.52) the acceleration can be calculated with Eq. (7.48).

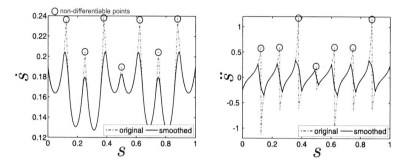

Figure 7.10: MVC and its time derivative for the circle path with ($|\dot{\tau}| \leq 1.6\ \frac{Nm}{s}$)

As obvious from Fig. 7.10, the MVC has non-differentiable points. These are to be smoothed applying the strategy introduced in Sec. 7.2.4. In order to do so the regions in which a smoothing is to be performed has to be selected first. In the $\dot{s}$-$\ddot{s}$ phase plane, cp. Fig. 7.10 (right), it can be seen, that $\ddot{s}$ at such a point is not continuous (jumping). The smoothing has been programmed in Matlab as a zero value search with event termination. The solver ode15s is used for the integration of the models. Applying the strategy for all points along the MVC it will be smoothed ($\mapsto$ step 1). According to Sec. 7.2.4 in a further step the transitions form the starting- and end-point of the path onto the MVC have to be determined. Again this problem is solved as an integration with event termination. Since the found transitions onto the MVC are in general not differentiable they have to be smoothed in a last step ($\mapsto$ step 2).

In Fig. 7.11 the original MVC and the PCSMTOPT in the $\dot{s}$-$\ddot{s}$ phase plane are shown exemplarily. The obtained PCSMTOPT for the circle path for a jerk limit of $|\dot{\tau}| \leq 1.6\ \frac{Nm}{s}$ is shown in Fig. C.4 – where on the left side the values of the pseudo variables and on the right side the corresponding physical values of the drives are given with respect to s. Planning the trajectories for less restricted jerks, less smooth trajectories are obtained, whereas the gains on the traveling time are only marginal, cp. Tab. 7.2 and App. C.2 (where all planned trajectories are summarized).

According to Prop. 1 the planned trajectories have a time-optimal characteristic. In case of a limitation of the jerk $\dot{\tau}$ also the pseudo-jerk $\dddot{s}$ is bounded. Furthermore, with Prop. 1 it can be concluded, that trajectories which are considering a limitation of the jerk are also smooth.

Especially in case of the triangle trajectory, cp. Fig. C.7 to C.9, the essential impact of the limitation of the jerk $\dot{\tau}$ on the trajectories becomes obvious. In case of a

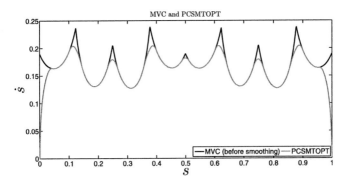

Figure 7.11: MVC and PCSMTOPT in the $\dot{s}$-$\ddot{s}$ ($|\dot{\tau}| \leq 1.6 \frac{\text{Nm}}{\text{s}}$)

jerk limit	circle	triangle	eight
$|\dot{\tau}| \leq 1.6 \frac{\text{Nm}}{\text{s}}$	6.5695 s	5.0888 s	–
$|\dot{\tau}| \leq 16 \frac{\text{Nm}}{\text{s}}$	6.2185 s	4.2934 s	–
$|\dot{\tau}| \to \infty$	6.1680 s	4.0999 s	3.5740 s

Table 7.2: Traveling times of the trajectories

limitation the transitions of $\ddot{s}$ are visibly smoothed. For that reason it is expected, that a strong limitation of the jerk, $|\dot{\tau}| \leq 1.6 \frac{\text{Nm}}{\text{s}}$ leads to a good performance in case of the trajectory tracking control in Chap. 10. Of further special interest are the transitions between the individual straights in case of the triangles – where $\dot{s}$ is braked to zero – and its impact on the control. In case of the eight-trajectory the transitions between the circular segments and the straights are of special interest, since for this case $\dot{s}$ is not continuous.

From the courses of the drive variables of all trajectories it can be seen, that at each path point s at least one of them is in its limitation. The only case where all of the three physical limits are reached are that of the triangle trajectory planned for a jerk limit of $|\dot{\tau}| \leq 16 \frac{\text{Nm}}{\text{s}}$. In order to emphasize this aspect all of the variable are normalized to their maximum value and plotted into one Fig. 7.12.

It has to be mentioned, that in case of the eight path it has not been possible to plan trajectories for limited jerks. There has always been a jump in the course of $\dot{s}$ at the transitions between the circle segments and the straights. Due to the high velocity along the straights and the related residual energy a smoothing is

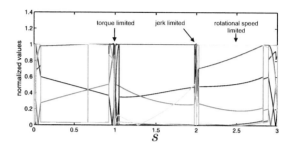

Figure 7.12: Normalized physical values of the drives [M08]

not possible and the solver shuts down. Further investigations are necessary.

7.4 Possible extensions of the strategy

7.4.1 Singular points and dynamic singularities

The problems planning trajectories for the manipulator SpiderMill in case of the eight-path considering jerk limits are related to the problem of dynamic singularities.

In order to give an idea of this aspect some of the results of Shiller and Lu [SL92, Shi94] are shortly summarized. The approach is very similar to the basic algorithm discussed in Sec. C.1.1. It works also in the phase space and only considers a limitation of the acceleration. Also here the time-optimality is guaranteed by the fact, that for each s at least one actuator is in its physical limit. In their work the authors show, that depending on the chosen path a time-optimal control has not to be necessarily of bang-bang type in its acceleration (or its corresponding 'chattering' actuator torques). The points on the MVC at which the acceleration is not in its limitation are called singular points if they are isolated and singular arcs if they are connected. In order to give a more detailed description of their influence, some of the results of Sec. C.1.1 are set in the new context.

The MVC is a combination of trajectory sinks and source with transition points between it. At a transition point, critical point or also called switching point there is a switch from a deceleration to a acceleration in case of a transition form a sink to a source. Whereas at a regular point on the MVC only a single value of the acceleration is admissible, cp. Fig. C.2 – case ⓐ and ⓒ, at a critical point the situation of case ⓑ occurs. In the following critical points have to be distinguished from singular ones. Whereas in a critical point the MVC gets in this point a part (tangent) of the time-optimal trajectory, while a limit of the acceleration is reached, in case of a singular point the maximum acceleration would drive the trajectory across the MVC. In order to compensate this the maximum feasible acceleration at a singular point is therefore selected to slide along the velocity constraint – which means along the MVC, with an acceleration which is neither the maximum nor the minimum acceleration – immediately after the singular point. In one sentence: in a singular point the PCTOPT does not maximize or minimize the acceleration along the defined path.

The effects of not considering singular points are among other things unbounded computation times, a chattering along the MVC or trajectories whose actuator torques cannot be smoothed. In [SL92] it is also emphasized, that algorithms using a cost function like that in Eq. (7.20) are very sensitive to the related effects. Furthermore a strategy for determining the singular points is introduced.

7.4.2 Trajectory planning in case of parallel robots considering friction

As already emphasized in Sec. 6.1.1 friction losses in the active and passive joints have to be considered in the dynamic model of a (parallel) robot to improve the quality of the planned trajectories and to enhance control performance in case of model-based control strategies.

For the case, that the friction losses of the joints are modeled as combination of Coulomb friction and viscous damping, the friction forces can be described by

$$\boldsymbol{\tau}_f = \mathbf{r}_V \, \dot{\mathbf{q}}_\mathbf{a} + \mathbf{r}_C \, \text{sign} \left(\dot{\mathbf{q}}_\mathbf{a} \right) \tag{7.54}$$

directly expressed in the space of the actuated joints. The matrices $\mathbf{r}_V$ and $\mathbf{r}_C$ are containing the corresponding friction coefficients. Based on Eq. (7.54) the consideration of not actuated joints in the trajectory planning becomes possible [AH05]. The changes of the signs of the passive joint velocities can be crucial [GAH04] since they lead to jumps in the dynamic forces due to dry friction. This discontinuities have a direct effect on the trajectory planning, since they have to be handled as further critical points.

Considering friction terms in the dynamic model Eq. (7.16) changes to [AH05]

$$\boldsymbol{\tau} = \mathbf{m} \left(s \right) \cdot \ddot{s} + \mathbf{c} \left(s \right) \cdot \dot{s}^2 + \mathbf{f} \left(s \right) \cdot \dot{s} + \mathbf{g}^* \left(s \right) \tag{7.55}$$

where the viscous friction extend the dynamics by the term $\mathbf{f} \left(s \right) \dot{s}$ and the dry friction is combined with $\mathbf{g}(s)$ to $\mathbf{g}^*(s)$. A closer analysis of Eq. (7.55) explains why in the most strategies viscous damping is neglected [SD85, PJ87, CC00]. The collective existence of the two terms $\mathbf{c}(s)\dot{s}^2$ and $\mathbf{f}(s)\dot{s}$ complicates the search for the maximum possible velocity, which now can no more be reduced to a linear problem in the $\dot{s}$-$\ddot{s}$ phase plane.

Results presented in [AH05] additionally show that even more the viscous friction terms seriously reduce the velocity and acceleration capacities of a parallel robot and not only the dry friction as often assumed. Also the assumption is made, that in case of parallel robots the existence of forbidden islands [SM85] within the admissible region is very probable. The neglect of the viscous friction in combination with forbidden islands can lead to a considerable control errors. As a further result it has been find out, that time phases, where more than one actuator torque reaches its bound are extremely short. This is a consequence of the highly coupled dynamics in case of parallel robots.

7.4.3 Adaptive PCSMTOPT

One further interesting possibility for the extension of the proposed strategy is the combination of the trajectory planning with an adaptive control approach as introduced by Pietsch [PBKH03, Pie03, PKB$^+$05, PBBH06]. This concept is useful when the dynamic parameters of robot are changing, which is a common situation in handling processes, where the payload is often varying and unknown. In order to handle this situation the parametrized inverse dynamics is set up in parameter-linear (and -minimal) form based on which the dynamic parameters can be identified, cp. Chap. 6. Applying the proposed strategy, a synergy effect occurs: Due to the fact that PC(SM)TOPT trajectories are planned and used for the first trajectory run – before the dynamic parameters become estimated and updated in the online trajectory planning for further cycles – the excitation of the dynamic parameters is normally sufficient for identification.

The performed planning of the time-optimal trajectories in the phase plane based on critical points is similar to other classical strategies, cp. Sec. C.1, whereas the reachable points on the MVC are detected by a new algorithm [Pie03]. Like the strategy introduced in Sec. 7.2 also the concept of Pietsch considers a limitation of the jerk in order to get smoother trajectories, but in a different way. The jerk limits are used in an implicit and an explicit manner: In the constraint definition part of the trajectory planning the jerk bounds are integrated with respect to time to redefine the acceleration bounds. This step is necessary, since the robot dynamics is formulated in an acceleration-separated form. But as a result of the proposed trajectory planning strategy at the intersection points of forward and backward trajectory segments high path jerks – violating the constraints – occur. In order to handle this problem these points are identified and an iterative jerk-limiting algorithm is applied until the limit is not longer violated. In case of the strategy presented in Sec. 7.2.4 such a step is not necessary, but from its effect it is a little bit similar to the smoothing procedure introduced above. A further problem of the strategy introduced in [PKB+05] is the fact that the trajectories temporarily exceeds the defined drive torque limits. Also here these limits are chosen lower than the physical possible values in order to guarantee enough reserve for the compensation of model errors and disturbances by the controller. The problem of the temporary violation of the torque constraints is handled with a procedure called online trajectory scaling [PKB+05]. Its basic idea is to limit the velocity at each path point s in a way, that the trajectory does not exceeds the defined limits. In order to do so a strategy proposed by Dahl [Dah92] has been adapted.

One of the inherent advantages of the strategy of Pietsch is the use of the parameter-linear and -minimal form of the inverse dynamics with its reduced computational effort. Furthermore it has to be mentioned that also in [Pie03, PKB+05] problems related with the eight path occur when the same limits for the dynamic parameters as in case of the circle and triangle path are chosen. The problems occur at the transitions between the straights and the circular arcs. At these points jumps in the acceleration occur in case of time-optimal planned trajectories. Also the torque bounds are much more violated compared with the other two path.

7.5 Conclusions and future works

In this chapter a new strategy for the planning of PCSMTOPTs in the phase space has been introduced and used to design trajectories for the SpiderMill. During the derivation some specific simplifications possible in case of the SpiderMill has been made and used to simplify the planning of the trajectories. A first extension of the strategy would be to avoid this simplifications, before in a further step a dynamic model including friction terms should be used. In this context the aspect of dynamic singularities should be analyzed in more detail and considered in the algorithm. Afterwards the concept of adaption can integrated.

8 Feedback linearization and stabilization

In robotics the well known concept of *inverse dynamics control* [SS05] is a special case of the static feedback linearization[1]. Whereas the latter one is collective term for a bundle of strategies. Their common goal is the linearization of special classes of nonlinear systems, which allows the application of standard linear control strategies in a further step of controller design.

The inverse dynamics control consists of two main parts: the feedback linearization [Isi89, SL91, Kha02, Mar03] and the stabilization. Altogether three steps are necessary: a state transformation, a linearization and a stabilization. First a state transformation is searched, which transforms the original nonlinear system in a new state space representation in which it is linearizable (step ⓐ). Based on an appropriate designed state feedback the nonlinearities of the system are compensated (step ⓑ). In contrast to a classical linearization [F94, DB05, FPEN06] the feedback linearization is not only exact in a working point, but rather in a defined region or in some special cases also globally. Two kinds of feedback linearization have to be distinguished the input-state (I/S) or exact feedback and the input-output-linearization (I/O). Whereas in the first case the interest is in linearizing the mapping from the inputs to the states, in case of tracking control problems – like here – it is rather in a certain output variable than in the state [Mar03]. Based on the I/O linearization a marginally stable, linear system is obtained which has to be stabilized in a further step (step ⓒ).

Two types of problems can arise in the context of feedback linearization: Singularities can occur in the feedback, limiting the area of validity of the linearization. Furthermore, the transformation of the states and the design of the state feedback are model-based. For that reason uncertainties of model-parameters and simplifications in the modeling process can lead to linearization errors.

In order to give a better impression in which direction the following steps will finally go Fig. 8.1 is given. After the end of this chapter the dashed box will be 'realized'. Defining one possible starting point for the realization of a trajectory tracking in Chap. 10.

---

[1]In the following the term feedback linearization is used to denote its static case.

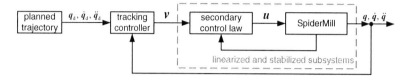

Figure 8.1: Concept of I/O-linearization in the context of trajectory tracking

The chapter is structured as follows: In Sec. 8.1 the feedback linearization problem is formulated and in Sec. 8.2 the concept is applied to the SpiderMill. The influence of uncertainties is analyzed in Sec. 8.3 and the stabilization of the linearized system is presented in Sec. 8.4. Conclusions are drawn in Sec. 8.5.

8.1 Problem formulation

In case of the I/S-linearization the mapping from the inputs to the states has to be linearized. But in case of trajectory tracking problems the output variables are of interest.

Given is the following input-affine system [Kha02]:

$$\dot{\mathbf{x}} = \mathbf{f}(\mathbf{x}) + \mathbf{g}(\mathbf{x}) \cdot \mathbf{u} \qquad\qquad \mathbf{f}, \mathbf{g} : D \subset \mathbb{R}^n \to \mathbb{R}^n \qquad (8.1a)$$

$$\mathbf{y} = \mathbf{h}(\mathbf{x}) \qquad\qquad\qquad \mathbf{h} : D \subset \mathbb{R}^n \to \mathbb{R}, \qquad (8.1b)$$

where $\mathbf{f}$, $\mathbf{g}$ and $\mathbf{h}$ are sufficiently smooth vector functions in a domain $D \subset \mathbb{R}^n$.
Linearizing its state equation (8.1a), in the form of an I/S-linearization problem, does not necessarily imply a linear mapping from the input $\mathbf{u}$ to the output $\mathbf{y}$. The reason for it is the fact, that while deriving the coordinate transformation $\phi(\mathbf{x})$ used to linearize the state equation (8.1a), the non-linearity in the output equation (8.1b) is not taken into account [Mar03].
In a short form the problem of the I/O-linearization can be formulated by the following questions [Kha02]: Exists a static state feedback control law

$$\mathbf{u} = \boldsymbol{\alpha}(\mathbf{x}) + \boldsymbol{\beta}(\mathbf{x}) \cdot \mathbf{v} \qquad (8.2)$$

and a change of variables (state transformation)

$$\mathbf{z} = \boldsymbol{\phi}(\mathbf{x}) \qquad (8.3)$$

that transforms the nonlinear system (8.1) into an equivalent linear and controllable system, between the new input $\mathbf{v}$ and the output $\mathbf{y}$?

The I/O-linearization is a standard procedure applied in robotics, where it is known under the term *inverse dynamics control*. But the derivation of linearized system can also be performed very formal following the steps defined in standard non-linear control literature [Isi89, SL91, Kha02, Mar03]. For that reason here not all definitions, which can be found in the mentioned books, are recapitulated. Rather the results in case of the SpiderMill are shortly summarized and set into a context to the definitions, where reasonable.

Problem 1 ([Isi89]) *Given is a nonlinear, input-affine system of the form* (8.1) *with the same number* m *of inputs and outputs, the corresponding vector fields* $\mathbf{f}(\mathbf{x})$ *and the matrix* $\mathbf{g}(\mathbf{x})$ *and an initial state* $\mathbf{x}^0$. *For exactly linearizing this system in a neighborhood* $\mathcal{N}$ *around point* $\mathbf{x}^0$ *with respect to its I/O behavior the feedback functions* $\boldsymbol{\alpha}(\mathbf{x})$ *and* $\boldsymbol{\beta}(\mathbf{x})$

$$\mathbf{u} = \boldsymbol{\alpha}(\mathbf{x}) + \boldsymbol{\beta}(\mathbf{x}) \cdot \mathbf{v} \quad \text{(linearizing feedback)} \tag{8.4}$$

and the state transformation

$$\mathbf{z} = \boldsymbol{\phi}(\mathbf{x}) \tag{8.5}$$

as well as the matrices $\mathbf{A} \in \mathbb{R}^{n \times n}$ *and* $\mathbf{B} \in \mathbb{R}^{n \times m}$ *have to be chosen that*

$$\left[\frac{\partial \boldsymbol{\phi}}{\partial \mathbf{x}} (\mathbf{f}(\mathbf{x}) + \mathbf{g}(\mathbf{x}) \cdot \boldsymbol{\alpha}(\mathbf{x})) \right]_{\mathbf{x} = \boldsymbol{\phi}^{-1}(\mathbf{z})} = \mathbf{A}\mathbf{z} \tag{8.6}$$

$$\left[\frac{\partial \boldsymbol{\phi}}{\partial \mathbf{x}} (\mathbf{g}(\mathbf{x}) \cdot \boldsymbol{\beta}(\mathbf{x})) \right]_{\mathbf{x} = \boldsymbol{\phi}^{-1}(\mathbf{z})} = \mathbf{B} \tag{8.7}$$

$$rank \begin{pmatrix} \mathbf{B} & \mathbf{A}\mathbf{B} & \dots & \mathbf{A}^{n-1}\mathbf{B} \end{pmatrix} = n. \tag{8.8}$$

with

$$\boldsymbol{\alpha}(\mathbf{x}) = \mathbf{A}^{-1}(\mathbf{x})\mathbf{B}(\mathbf{x}) \quad and \quad \boldsymbol{\beta}(\mathbf{x}) = \mathbf{A}^{-1}(\mathbf{x}) \tag{8.9}$$

and

$$\mathbf{A}(\mathbf{x}) = \begin{bmatrix} L_{g_1} L_f^{r_1-1} h_1(\mathbf{x}) & \dots & L_{g_m} L_f^{r_1-1} h_1(\mathbf{x}) \\ \vdots & & \vdots \\ L_{g_1} L_f^{r_m-1} h_m(\mathbf{x}) & \dots & L_{g_m} L_f^{r_m-1} h_m(\mathbf{x}) \end{bmatrix}, \mathbf{B}(\mathbf{x}) = \begin{bmatrix} L_f^{r_1} h_1(\mathbf{x}) \\ L_f^{r_2} h_2(\mathbf{x}) \\ \vdots \\ L_f^{r_m} h_m(\mathbf{x}) \end{bmatrix} \tag{8.10}$$

The derivation of the transformation matrix can be found in [Isi89].
In order to introduce the basic idea in an intuitive way in the following the SISO case of (8.1) is studied analogous to [Kha02]: The time-derivative of y is given by

$$\dot{y} = \frac{\partial h}{\partial \mathbf{x}} \left[\mathbf{f}(\mathbf{x}) + \mathbf{g}(\mathbf{x})u \right] \overset{\text{Lie}}{=} L_f h(\mathbf{x}) + L_g h(\mathbf{x})u \qquad (8.11)$$

using the Lie notation. If $L_g h(\mathbf{x}) = 0$, then $\dot{y}$ is independent of the input u. Calculating the second derivative $y^{(2)}$ of y

$$y^{(2)} = \frac{\partial L_f h}{\partial \mathbf{x}} \left[\mathbf{f}(\mathbf{x}) + \mathbf{g}(\mathbf{x})u \right] = L_f^2 h(\mathbf{x}) + L_g L_f h(\mathbf{x})u \qquad (8.12)$$

is obtained. If $L_g L_f h(\mathbf{x}) = 0$, then $y^{(2)} = L_f^2 h(\mathbf{x})$ is again independent of u. This process is repeated until

$$L_g L_f^{i-1} h(\mathbf{x}) = 0, \; i = 1, 2, \ldots, r-1; \quad L_g L_f^{r-1} h(\mathbf{x}) \neq 0 \qquad (8.13)$$

holds. Even though u does not appear in the equations of $y, \dot{y}, \ldots, y^{(r-1)}$ it appears in that of $y^{(r)}$ with nonzero coefficient

$$y^{(r)} = L_f^r h(\mathbf{x}) + L_g L_f^{r-1} h(\mathbf{x})u. \qquad (8.14)$$

Equation (8.14) indicates that the system is I/O-linearizable, since the state feedback control law

$$u = \frac{1}{L_g L_f^{r-1} h(\mathbf{x})} \left[-L_f^r h(\mathbf{x}) + v \right] \qquad (8.15)$$

reduces the mapping of the input to the output to a chain of r integrators

$$y^{(r)} = v. \qquad (8.16)$$

The integer r is called *relative degree* of the system [Kha02].

8.2 I/O linearization of the SpiderMill

The I/O linearization can be applied to each mechanical system, which can be expressed in some kind of Lagrangian form (5.18) (cp. e.g. [SK08, SS05]) and consequently also to the SpiderMill. For that reason the method is directly applied to the input-affine state-space model (5.28) of the SpiderMill. In general, applying this approach to a parallel robot (decoupled) integrator chains are obtained. In case of the SpiderMill the relative degrees of the subsystems r_i can be determined to $r_1 = r_2 = 2$.

According to Prob. 1, for the existence of the linearizing feedback it has to be guaranteed, that the matrix $\mathbf{A}(\mathbf{x})$ is invertible. Since in case of the SpiderMill

$$\mathbf{A}(\mathbf{x}) = \begin{bmatrix} L_{g_1} L_f^1 h_1 & L_{g_2} L_f^1 h_1 \\ L_{g_1} L_f^1 h_2 & L_{g_2} L_f^1 h_2 \end{bmatrix} = \mathbf{M}^{-1} \tag{8.17}$$

holds, the mass matrix $\mathbf{M}$ has to be checked for singularities, which do not occur. Therefore the state-space model of the SpiderMill has in the whole workspace a relative degree of $r_1 = r_2 = 2$, which implies, that each output has to be differentiated two times until one of the input-variable occurs for the first time. Furthermore, since the relative degree r is equal to the order n of the system

$$r = r_1 + r_2 = 4 = n \tag{8.18}$$

the model (5.28) is both I/O- and I/S-linearizable.

Below the results of the first two steps of the I/O-linearization are summarized.

ⓐ **State transformation**: First a state transformation $\boldsymbol{\phi}(\mathbf{x})$ is searched, which transforms the original nonlinear system in a new state space representation (state variable $\mathbf{z}$) in which it is linearizable. In case of the SpiderMill the transformation rule can be determined to

$$\mathbf{z} = \begin{bmatrix} z_1^1 \\ z_2^1 \\ z_1^2 \\ z_2^2 \end{bmatrix} = \begin{bmatrix} x_1 \\ x_3 \\ x_2 \\ x_4 \end{bmatrix} = \begin{bmatrix} 1 & 0 & 0 & 0 \\ 0 & 0 & 1 & 0 \\ 0 & 1 & 0 & 0 \\ 0 & 0 & 0 & 1 \end{bmatrix} \cdot \mathbf{x} = \mathbf{T} \cdot \mathbf{x} = \boldsymbol{\phi}(\mathbf{x}). \tag{8.19}$$

ⓑ **Linearizing feedback**: By differentiating the transformation rule (8.19) with respect to time and inserting the states according to Eq. (5.28) the so-called *Byrnes-Isidori normal form* is obtained

$$\begin{bmatrix} \dot{z}_1^1 \\ \dot{z}_2^1 \\ \dot{z}_1^2 \\ \dot{z}_2^2 \end{bmatrix} = \begin{bmatrix} z_2^1 \\ [-\mathbf{M}^{-1}(\mathbf{C}\dot{\mathbf{q}}^2 + \mathbf{G} + \boldsymbol{\tau}_f)]_{(1)} \\ z_2^2 \\ [-\mathbf{M}^{-1}(\mathbf{C}\dot{\mathbf{q}}^2 + \mathbf{G} + \boldsymbol{\tau}_f)]_{(2)} \end{bmatrix} + \begin{bmatrix} 0 & 0 \\ [\mathbf{M}^{-1}]_{(1,1)} & [\mathbf{M}^{-1}]_{(1,2)} \\ 0 & 0 \\ [\mathbf{M}^{-1}]_{(2,1)} & [\mathbf{M}^{-1}]_{(2,2)} \end{bmatrix} \cdot \begin{bmatrix} u_1 \\ u_2 \end{bmatrix}$$

$$\begin{bmatrix} y_1 \\ y_2 \end{bmatrix} = \begin{bmatrix} z_1^1 \\ z_1^2 \end{bmatrix}. \tag{8.20}$$

According to the requirements above, this state-space model is linearizable by state feedback. Since $r = n$ no internal dynamics exist in case of the SpiderMill and has therefore not to be checked for stability.

The feedback can according to Eq. (8.20) be determined to

$$\mathbf{u} = \mathbf{M} \cdot \mathbf{v} + \mathbf{C} \cdot \dot{\mathbf{q}}^2 + \mathbf{G} + \boldsymbol{\tau}_f, \tag{8.21}$$

yielding the following state-space models

$$\dot{\mathbf{z}}^i = \underbrace{\begin{bmatrix} 0 & 1 \\ 0 & 0 \end{bmatrix}}_{\mathbf{A}_{i,z}} \cdot \mathbf{z}^i + \underbrace{\begin{bmatrix} 0 \\ 1 \end{bmatrix}}_{\mathbf{B}_{i,z}} \cdot v_i \qquad y_i = \underbrace{\begin{bmatrix} 1 & 0 \end{bmatrix}}_{\mathbf{C}_{i,z}} \cdot \mathbf{z}^i \qquad \text{with} \quad \mathbf{z}^i = \begin{bmatrix} z_1^i \\ z_2^i \end{bmatrix} \quad \text{for} \quad i = 1, 2$$

$$\text{initial state:} \quad \mathbf{z}_0^i = \begin{bmatrix} z_{1,0}^i & z_{2,0}^i \end{bmatrix}^T. \tag{8.22}$$

in *Brunowsky canonical form*, a general result in robotics [SHV06].

The overall system consists of two decoupled subsystems, whereas both are double integrators. Thus, the I/O behavior between the input v_i and the corresponding output z^i is linear. All conditions of Prob. 1 are fulfilled. As a consequence the linearized system is controllable. The overall system is given by

$$\dot{\mathbf{z}} = \underbrace{\begin{bmatrix} \mathbf{A}_{1,z} & 0 \\ 0 & \mathbf{A}_{2,z} \end{bmatrix}}_{\mathbf{A}_z} \cdot \mathbf{z} + \underbrace{\begin{bmatrix} \mathbf{B}_{1,z} & 0 \\ 0 & \mathbf{B}_{2,z} \end{bmatrix}}_{\mathbf{B}_z} \cdot \mathbf{v}, \qquad \mathbf{y} = \underbrace{\begin{bmatrix} \mathbf{C}_{1,z} & 0 \\ 0 & \mathbf{C}_{2,z} \end{bmatrix}}_{\mathbf{C}_z} \cdot \mathbf{z}$$

$$\text{with} \quad \mathbf{z} = \begin{bmatrix} \mathbf{z}^1 & \mathbf{z}^2 \end{bmatrix}^T, \qquad \mathbf{v} = \begin{bmatrix} v_1 & v_2 \end{bmatrix}^T, \qquad \mathbf{y} = \begin{bmatrix} y_1 & y_2 \end{bmatrix}^T. \tag{8.23}$$

In order to verify the results of the feedback linearization the SimMechanics model of the SpiderMill is used, neglecting the influence of friction in the derivation above. Since the applied feedback (8.21) uses matrices of the reduced Newton-Euler model – not exactly reproducing the system dynamics – some linearization error will occur. In order to evaluate its effect the following linear trajectories are implemented

$$s_k(t) = s_l(t) = s_r(t) = 0.02\, t + 0.4672. \tag{8.24}$$

As reference systems two double integrator chains are used. The results are shown in Fig. 8.2. It is obvious, that the behavior of the I/O-linearized system and the reference system are similar, but not equivalent. A perfect I/O-linearization and decoupling based on a feedback using the matrices of the reduced Newton-Euler model cannot be achieved. But the feedback linearized model can and will be used for observer and controller design in the next chapters. To understand this point better, the influence of the linearization error is analyzed in the following.

---

[2]In case of asymmetric profiles the derivations between linearized and reference system are in the same range.

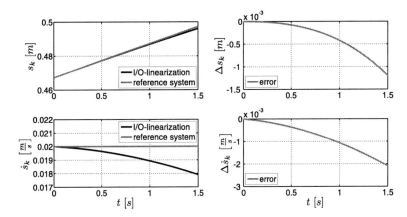

Figure 8.2: Verification of the linearization[2]

8.3 Influence of uncertainties

The derivation above assumes that the model of the robot is exact. This does not hold in practice, where the model parameters are only known with some uncertainty (even after parameter-identification) or even are changing (e.g. the end-effector mass in handling processes or friction parameters). Also in case of the SpiderMill this holds. Its model including the uncertainties is given by

$$\boldsymbol{\tau} = \hat{\mathbf{M}}\left(\mathbf{q}\right) \cdot \ddot{\mathbf{q}}_a + \hat{\mathbf{C}}\left(\mathbf{q}\right) \cdot \dot{\mathbf{q}}^2 + \hat{\mathbf{G}}\left(\mathbf{q}\right) + \hat{\boldsymbol{\tau}}_f\left(\dot{\mathbf{q}}\right). \tag{8.25}$$

where the $(\hat{\cdot})$ notation indicates the uncertainties. Furthermore it is assumed, that the SimMechanics model describes the physical robot 'exact' (with only marginal error).

Despite the small modeling errors the inverse dynamics (8.25) is used as linearizing feedback (8.21)

$$\mathbf{u} = \hat{\mathbf{M}}\left(\mathbf{q}\right) \cdot \mathbf{v} + \hat{\mathbf{C}}\left(\mathbf{q}\right) \cdot \dot{\mathbf{q}}^2 + \hat{\mathbf{G}}\left(\mathbf{q}\right) + \hat{\boldsymbol{\tau}}_f\left(\dot{\mathbf{q}}\right). \tag{8.26}$$

Inserting Eq. (8.26) in (8.20) yields [SS05, SHV06]

$$\dot{\mathbf{z}} = \mathbf{v} - \mathbf{M}^{-1} \cdot \left(\tilde{\mathbf{M}} \cdot \mathbf{v} + \tilde{\mathbf{C}} \cdot \dot{\mathbf{q}}^2 + \tilde{\mathbf{G}} + \tilde{\boldsymbol{\tau}}_f\right) \tag{8.27}$$

with $(\tilde{\cdot}) = (\cdot) - (\hat{\cdot})$. A comparison of Eq. (8.27) with Eq. (8.22) shows, that due to the uncertainties in the state feedback (8.26) the overall system is no more linearized in terms of its I/O behavior. It is affected by the uncertainty $\boldsymbol{\eta}$

$$\boldsymbol{\eta}\left(\mathbf{q}, \dot{\mathbf{q}}, \mathbf{v}\right) = \mathbf{M}^{-1} \cdot \left(\tilde{\mathbf{M}} \cdot \mathbf{v} + \tilde{\mathbf{C}} \cdot \dot{\mathbf{q}}^2 + \tilde{\mathbf{G}} + \tilde{\boldsymbol{\tau}}_f\right). \tag{8.28}$$

Due to that reason linear and decoupled subsystems are not obtained via the feedback linearization. The subsystem model (8.22) changes to

$$\dot{\mathbf{z}}^i = \underbrace{\begin{bmatrix} 0 & 1 \\ 0 & 0 \end{bmatrix}}_{\mathbf{A}_{i,z}} \cdot \mathbf{z}^i + \underbrace{\begin{bmatrix} 0 \\ 1 \end{bmatrix}}_{\mathbf{B}_{i,z}} \cdot (v_i - \eta_i) \quad y_i = \underbrace{\begin{bmatrix} 1 & 0 \end{bmatrix}}_{\mathbf{C}_{i,z}} \cdot \mathbf{z}^i \quad \text{with } \mathbf{z}^i = \begin{bmatrix} z_1^i \\ z_2^i \end{bmatrix} \quad \text{and } i = 1, 2$$

$$\text{initial state:} \quad \mathbf{z}_0^i = \begin{bmatrix} z_{1,0}^i & z_{2,0}^i \end{bmatrix}^T$$

$$\tag{8.29}$$

and consequently the overall model to

$$\dot{\mathbf{z}} = \underbrace{\begin{bmatrix} \mathbf{A}_{1,z} & 0 \\ 0 & \mathbf{A}_{2,z} \end{bmatrix}}_{\mathbf{A}_z} \cdot \mathbf{z} + \underbrace{\begin{bmatrix} \mathbf{B}_{1,z} & 0 \\ 0 & \mathbf{B}_{2,z} \end{bmatrix}}_{\mathbf{B}_z} \cdot (\mathbf{v} - \boldsymbol{\eta}) \qquad \mathbf{y} = \underbrace{\begin{bmatrix} \mathbf{C}_{1,z} & 0 \\ 0 & \mathbf{C}_{2,z} \end{bmatrix}}_{\mathbf{C}_z} \cdot \mathbf{z}$$

$$\text{with} \qquad \mathbf{z} = \begin{bmatrix} \mathbf{z}^1 \\ \mathbf{z}^2 \end{bmatrix}, \qquad \mathbf{v} = \begin{bmatrix} v_1 \\ v_2 \end{bmatrix}, \qquad \mathbf{y} = \begin{bmatrix} y_1 \\ y_2 \end{bmatrix}, \qquad \boldsymbol{\eta} = \begin{bmatrix} \eta_1 \\ \eta_2 \end{bmatrix}. \tag{8.30}$$

The uncertainty can be interpreted as input-disturbance introducing some kind of coupling of the two subsystems and impressing a noticeable nonlinear characteristic to the overall system, cp. Fig. 8.3. Its influence becomes obvious from Fig. 8.2. Nevertheless, the feedback linearization can be used for the control of the Spider-Mill. Despite the uncertainty, the overall systems will be regarded as linear, decoupled system, consisting of two double integrators for controller design. Thereby, the uncertainty $\boldsymbol{\eta}$ will be handled as disturbance at the system input.

8.4 Stabilization of the linearized system

The last step after the linearization is the design of a secondary control law, stabilizing the system and imposing the desired performance [Mar03].

© **Secondary control law**: Analyzing $\mathbf{A}_{i,z}$ it is obvious, that the subsystem i has two eigenvalues in the origin and is marginally stable. Thus it has to be stabilized in order to be used within a closed-loop control. Furthermore the system dynamics should be enhanced with respect to the trajectory tracking control in Chap. 10.

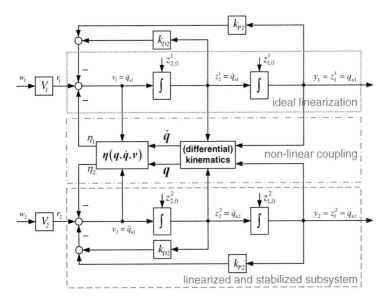

Figure 8.3: Feedback linearization with uncertainty and stabilization

The control laws for both subsystems are designed identical.
A simple state feedback with a constant matrix $\mathbf{K}_i\,(t, \mathbf{z}^i)$ is sufficient

$$v_i = -\mathbf{K}_i \cdot \mathbf{z}^i + r_i = -\begin{bmatrix} k_{Pi} & k_{Di} \end{bmatrix} \cdot \begin{bmatrix} z_1^i \\ z_2^i \end{bmatrix} + r_i. \qquad (8.31)$$

At this point it is assumed, that all states are accessible. The state z_1^i is fed back via k_{Pi} and z_2^i via k_{Di} to the input v_i. Additionally, with r_i a new system input is defined, which is necessary for the control of the system. The procedure is shown schematically in Fig. 8.3. Inserting Eq. (8.31) into (8.29) yields

$$\begin{bmatrix} \dot{z}_1^i \\ \dot{z}_2^i \end{bmatrix} = \begin{bmatrix} 0 & 1 \\ -k_{Pi} & -k_{Di} \end{bmatrix} \cdot \begin{bmatrix} z_1^i \\ z_2^i \end{bmatrix} + \begin{bmatrix} 0 \\ 1 \end{bmatrix} \cdot (r_i - \eta_i) \qquad y_i = \begin{bmatrix} 1 & 0 \end{bmatrix} \cdot \mathbf{z}^i. \qquad (8.32)$$

By the choice of k_{Pi} and k_{Di} the dynamics of the second order system can be influenced. For the choice of $\mathbf{K}_i$ standard requests are applied in the following [DB05, FPEN06]. This means that beside a pure stabilization and the compensation of initial displacements (of $z_{1,0}^i$ and $z_{2,0}^i$) the system-output should tend to the reference value w_i for $t \to \infty$. Whereas the first two point are request on the gains itself, the last condition can be reached by a *prefilter* [F94, DB05]

$$r_i = V_i \cdot w_i. \tag{8.33}$$

Its design in case of trajectory tracking control is concretized in Sec. 10.3.

For analyzing the transmission behavior with respect to a disturbance, the system input r_i is set to zero, yielding the transfer function G_{si} between y_i and η_i

$$G_{si}(s) = \frac{Y_i(s)}{\eta_i(s)} = \frac{-1}{s^2 + k_{Di} \cdot s + k_{Pi}}. \tag{8.34}$$

From Eq. (8.34) it is obvious, that the influence of the uncertainty decreases with increasing controller gains. For analyzing exemplarily the influence of η_i on the stabilized system (8.32), the steady state value of the output $Y_i(s)$ is determined for a unit step $\eta_i(s) = \frac{1}{s}$ as disturbance

$$\lim_{t \to \infty} y_i(t) = \lim_{s \to 0} s \cdot G_{si}(s) \cdot \eta_i(s) = -\frac{1}{k_{Pi}}. \tag{8.35}$$

For that case the steady state error depends only on k_{Pi}. As obvious from Fig. 8.4 a higher value of k_{Pi} leads to a faster convergence, but produces at the same time also higher 'amplitudes' (since deviations are multiplied with higher gains – also k_{Di} has an influence on the transient behavior). The matrix $\mathbf{K}_i$ has to be chosen in order to minimize the influence of $\boldsymbol{\eta}$ while at the same time enhancing the system dynamics, since the stabilized system is – in some cases – part of the trajectory tracking control in Chap. 10.

The compensation of initial displacements and the reaching of the reference value should be done with a small rise time t_r (10% to 90%) and without overshooting. A upper limit for t_r is chosen to $t_r < 0.05$ s. In order to fulfill the defined specifications two sets of controller parameters are introduced. They and the related eigenvalues of the overall system are summarized in Tab. 8.1. The reason why two sets are introduced becomes obvious in Sec. 10.3.2.

For testing the behavior of the linear system (8.22) for the two gain matrices

controller	gains		eigenvalues		rise time
C1	$k_{Pi} = 80500$	$k_{Di} = 580$	$\lambda_1^i = -230$	$\lambda_2^i = -350$	13 ms
C2	$k_{Pi} = 6000$	$k_{Di} = 160$	$\lambda_1^i = -60$	$\lambda_2^i = -100$	46 ms

Table 8.1: Gains, eigenvalues and rise times

two situations are analyzed: The compensation of initial state errors $z_{1,0}^i = 1$ and

$z^i_{2,0} = 0.5$ as well as the reaction of the system output on an unit step as reference input w_i. In the last case a prefilter V_i has to be designed according to Eq. (8.33) [F94]

$$V_i = k_{Pi}. \tag{8.36}$$

The results are shown in Fig. 8.4. The initial values $z^i_{1,0}$ and $z^i_{2,0}$ are compensated

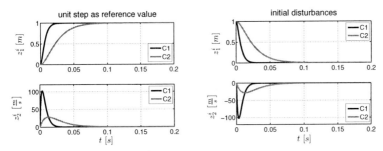

Figure 8.4: Behavior of the linear, stabilized reference system for the test cases

and the reference value for the spindle length w_i is tracked. Both without overshooting and with a rise time below $t_r < 0.05$ s. Due to the high gains, especially in case of C1 it can be expected, that the influence of the uncertainty η on the stabilized system is only marginal.

In a second step the SimMechanics model of the robot in combination with the linearizing feedback (8.21) and the stabilization (8.31) is implemented. The resulting overall structure is sketched in Fig. 8.5, where $\mathbf{K} = [\mathbf{K}_1, \ \mathbf{K}_2]^T$.

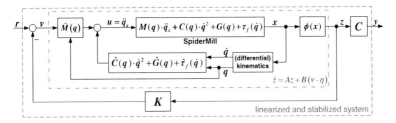

Figure 8.5: Feedback linearization with uncertainty and stabilizing state feedback[3]

---

[3]The block *(differential) kinematics* is not the inverse or direct kinematics of the robot. It describes the relationships for calculating the passive coordinates and their time-derivatives.

For better analyzing the influence of $\boldsymbol{\eta}$, the system behavior is compared to the ideal case, using the linear system (8.22) as reference. For the comparison the following test scenario is defined: for both subsystems an initial displacement has to be compensated and a defined final value $y_{i,f}$ to be reached. In order to allow a better interpretation of the values the initial spindle lengths $s_{l,0} = z_{1,0}^1$ and $s_{r,0} = z_{1,0}^2$ are that of the starting point of the circle path and the final lengths $s_{l,f} = y_{1,f}$ and $s_{r,f} = y_{2,f}$ are that after one quadrant.

subsystem 1: $z_{1,0}^1 = 0.5741$ m $z_{2,0}^1 = 0.0\,\dfrac{\text{m}}{\text{s}}$ $y_{1,f} = 0.5858$ m   (8.37a)

subsystem 2: $z_{1,0}^2 = 0.5858$ m $z_{2,0}^2 = 0.0\,\dfrac{\text{m}}{\text{s}}$ $y_{2,f} = 0.5279$ m   (8.37b)

For the reaching of the final value again the prefilter has to be chosen according to Eq. (8.36). The results are shown in Fig. 8.6 and 8.7. The behavior of the

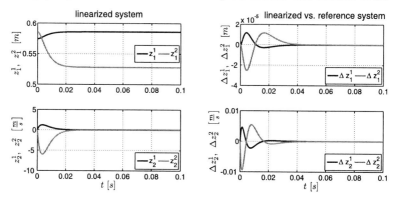

Figure 8.6: Linearized and reference system for C1

linearized and stabilized systems are almost identical to the reference system, which is obvious from the error plots. The effect of the the uncertainty is only minimal. The errors converge in case of C1 faster, but also with slightly larger amplitudes. The state feedback with both gain matrices satisfy all requests. Consequently the stabilized system (8.32) can be used as part of the trajectory-tracking control.

8.5 Conclusions and future works

In this chapter the feedback linearization has been performed for the reduced Newton-Euler model of the SpiderMill. Thereby the unavoidable uncertainties in

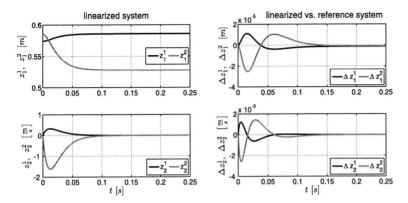

Figure 8.7: Linearized and reference system for C2

case of an analytical modeling have been considered. Keeping the goal of trajectory tracking control in mind, a fast dynamics of the linearized and stabilized system has been guaranteed. The obtained overall system defines the basis for the reference trajectory tracking control in Sec. 10.3. An interesting further point would be to analyze the other analytical models derived in Chap. 5 with respect to their uncertainties in case of feedback linearization.

.

9 Velocity reconstruction

Many control approaches, e.g. the state-space control, require information about all system states. Normally this information is not accessible by direct measurement – the effort is too large or there is too much measurement noise. In robotics, in case of trajectory tracking control the missing information is normally that of the velocities and sometimes that of the accelerations. In case of the control concepts of Chap. 10 the required information is that of spindle (or angular) velocities. Two principle ways to get the missing information have to be distinguished. The first one is model based. The missing or even all system states are determined by observer or filter structures. The other class of strategies does not require a system model and bases on a kind of time derivative of the position signal. Strictly speaking, the term observer is only correct for the first class of methods, nevertheless for simplicity for all strategies the term *observation* is used. Both classes of strategies will be discussed for the SpiderMill in this chapter.

In case of the model-based concepts the feedback-linearized and stabilized system derived in Chap. 8 is the starting point. This implies that the observer is designed in the state space of $\mathbf{z}$. In order to realize the linearization and stabilization the entire-state vector $\mathbf{x}$ is required. As obvious form Eq. (5.28) only a part of it (the spindle length) can be measured at the system output. For the reconstruction of the entire state vector first a High Gain Observer (HGO) and afterwards a Linear Kalman-Filter (LKF)[1] are applied. In both cases two identical observers are designed for the two linear double-integrator subsystems representing the linearized manipulator dynamics.

The overall question for the second class of strategies is: Is it not possible to simply build the finite difference of the encoder signals? In order to discuss this aspect the analysis performed in [BDHZ98] is applied to the encoders of the SpiderMill neglecting the influence of measurement noise. The encoders have an interpulse angle θ_m and are sampled with the sampling rate T_S of the control system. A number of N_p pulses occurs during the interval T_S. The angular velocity can be estimated by the finite difference $N_p \cdot \theta_m / T_S$, implying that the estimation is quantized with a step-size of θ_m / T_S. Since the most control strategies in Chap. 10 are based on the control of the spindle lengths and velocities (or can be implemented

---

[1] In this work the term Kalman-Filter is used to refer to its discrete version.

in this form) the analysis has to be adapted for the ball-screw systems in combination with the motor-shaft encoders. In case of a full revolution the spindle lengths change by $\delta s = h = 5 \cdot 10^{-3}$ m. This yields and interpulse length change of $\theta_m^\star = 0.0012 \cdot 10^{-3}$ m (encoder resolution: 1024 pulses/revolution × 4 (quadrature mode)) and for $T_S = 0.001$ s: $\theta_m^\star / T_S = 1.22 \cdot 10^{-3}$ m/s. This resolution is much better than those obtained in case of direct measurements of the angles of the active axis and would in this special case also allow a numerical differentiation of the position signals in the absence of too intense noise – cp. Sec. 10.6. But ball screw systems are not standard in robotics. A typical[2] $\theta_m^{(\star)} / T_S = 87.89°/$s does not allow numerical differentiation, especially in case of higher sampling rates or when also the acceleration $(\theta_m^{(\star)} / T_S^2)$ ought to be determined. For that reason the more sophisticated Adaptive Windowing (AW) is discussed below.

This chapter is structured as follows: In Sec. 9.1 the theory of the HGO is introduced and HGOs for the SpiderMill are designed. The same is done in Sec. 9.2 for the LKF. The idea of AW and two concrete algorithms are introduced in Sec. 9.3. Conclusions are drawn in Sec. 9.4.

9.1 High Gain Observer (HGO)

The High Gain Observer (HGO) [GHO92] can be used for a wide class of non-linear systems. Due to the high gain in its correction term it is outstandingly robust against model uncertainties. Since the feedback linearization of a physical system is normally not exact (cp. Sec. 8.3), this aspect plays an important role in practice.

In this section the structure of the HGO will be derived leaning on [Kha02]. The observer will be used for the estimation of the system states, which will be used in the feedback linearization and in the stabilizing state feedback.

9.1.1 Theory of the HGO

The structure of the HGO is derived for a non-linear system of the form [Kha02]

$$\dot{\mathbf{x}} = \mathbf{A}_x \cdot \mathbf{x} + \boldsymbol{\varphi}_x \left(\mathbf{y}, \mathbf{u} \right) \tag{9.1a}$$

$$\mathbf{y} = \mathbf{C}_x \cdot \mathbf{x} \tag{9.1b}$$

where $\mathbf{y}$ is the measured output, $(\mathbf{A}_x, \mathbf{C}_x)$ is observable and $\boldsymbol{\varphi}_x \left(\mathbf{y}, \mathbf{u} \right)$ is a non-linear part. The index x indicates that the system is expressed in the state space of

---

[2]The notation $(\star)$ is used to distinguish between lengths and angles.

$\mathbf{x}$. The form of the model in Eq. (9.1) is special, since the non-linear part depends only on the system output $\mathbf{y}$ and the control input $\mathbf{u}$. For this class of non-linear systems an observer of the form

$$\dot{\hat{\mathbf{x}}} = \mathbf{A}_x \cdot \hat{\mathbf{x}} + \boldsymbol{\varphi}_x (\mathbf{y}, \mathbf{u}) + \mathbf{H}_x (\mathbf{y} - \mathbf{C}_x \hat{\mathbf{x}}) \qquad (9.2)$$

can be introduced, where $\mathbf{H}_x$ is the estimator gain matrix. The observation error $\tilde{\mathbf{x}} = \mathbf{x} - \hat{\mathbf{x}}$ satisfies the following linear equation

$$\dot{\tilde{\mathbf{x}}} = (\mathbf{A}_x - \mathbf{H}_x \cdot \mathbf{C}_x) \tilde{\mathbf{x}}, \qquad (9.3)$$

which allows the design of a linear observer. But since in practice the non-linear function $\boldsymbol{\varphi}_x$ is not perfectly known a further analysis is performed in [Kha02]. Shortly summarized, the result is that for large (see below) controller gains the effect of a not perfectly known $\boldsymbol{\varphi}_x$ vanishes for the limit case.

In [GHO92, BZH02] the HGO design for a system of the structure

$$\dot{\mathbf{x}} = \mathbf{A}_{0x} \cdot \mathbf{x} + \boldsymbol{\varphi}_{0x} (\mathbf{x}, \mathbf{u}) \qquad (9.4\text{a})$$

$$y = \mathbf{C}_{0x} \cdot \mathbf{x}, \qquad (9.4\text{b})$$

where

$$\mathbf{A}_{0x} = \begin{bmatrix} 0 & 1 & 0 & \dots & 0 \\ \vdots & & \ddots & & \vdots \\ & & & & 1 \\ 0 & & \dots & & 0 \end{bmatrix}, \quad \mathbf{C}_{0x} = \begin{bmatrix} 1 \\ 0 \\ \vdots \\ 0 \end{bmatrix}^T, \quad \boldsymbol{\varphi}_{0x} = \begin{bmatrix} \varphi_{0x,1}(x_1, \mathbf{u}) \\ \varphi_{0x,2}(x_1, x_2, \mathbf{u}) \\ \vdots \\ \varphi_{x,n}(\mathbf{x}, \mathbf{u}) \end{bmatrix}$$

and $\mathbf{x} = [x_1, x_2, \dots x_n]^T$ is presented and in [BH98] extended to a system with two outputs

$$\dot{\mathbf{x}} = \underbrace{\begin{bmatrix} \mathbf{A}_{1x} & 0 \\ 0 & \mathbf{A}_{2x} \end{bmatrix}}_{\mathbf{A}_x} \cdot \mathbf{x} + \underbrace{\begin{bmatrix} \boldsymbol{\varphi}_{1x} (\mathbf{x}, \mathbf{u}) \\ \boldsymbol{\varphi}_{2x} (\mathbf{x}, \mathbf{u}) \end{bmatrix}}_{\boldsymbol{\varphi}_x} \qquad (9.5\text{a})$$

$$\mathbf{y} = \underbrace{\begin{bmatrix} \mathbf{C}_{1x} & 0 \\ 0 & \mathbf{C}_{2x} \end{bmatrix}}_{\mathbf{C}_x} \cdot \mathbf{x}, \qquad (9.5\text{b})$$

where $\mathbf{A}_{1x}$, $\mathbf{A}_{2x}$ and $\mathbf{C}_{1x}$, $\mathbf{C}_{2x}$ are of the same form like $\mathbf{A}_{0x}$ and $\mathbf{C}_{0x}$. For the vector $\mathbf{x} = [\mathbf{x}[1], \mathbf{x}[2]]^T$ holds, where $\mathbf{x}[i]$ is of the same dimension as $\mathbf{A}_{ix}$. The functions $\boldsymbol{\varphi}_{1x}$ and $\boldsymbol{\varphi}_{2x}$ are of the same structure as $\boldsymbol{\varphi}_{0x}$, but their respective last components can additionally depend on $\mathbf{x}[j]$ with $j \neq i$ [BZH02].

The input-affine model of the SpiderMill (5.28) can only be expressed by

$$\dot{\mathbf{x}} = \mathbf{S}_x \cdot \mathbf{x} + \boldsymbol{\varphi}_{sx}(\mathbf{x}, \mathbf{u}) \qquad \text{with} \quad \boldsymbol{\varphi}_{sx} = \begin{bmatrix} \mathbf{0}^T & \boldsymbol{\varphi}_{sx}^T \end{bmatrix}^T, \tag{9.6}$$

with $\mathbf{S}_x$ having not the form of $\mathbf{A}_x$ in Eq. (9.5a). Applying a not further specified non-linear observer, the dynamics of the closed-loop is given by

$$\dot{\hat{\mathbf{x}}} = \mathbf{S}_x \cdot \hat{\mathbf{x}} + \hat{\boldsymbol{\varphi}}_x(\hat{\mathbf{x}}, \mathbf{u}) + \mathbf{L}_x \cdot (\mathbf{y} - \mathbf{C}_x \cdot \hat{\mathbf{x}}) \qquad \text{(n.l. observer)}, \tag{9.7}$$

where $\mathbf{L}_x$ is the estimator gain matrix. As dynamics of the observation error

$$\dot{\tilde{\mathbf{x}}} = (\mathbf{S}_x - \mathbf{L}_x \cdot \mathbf{C}_x)\,\tilde{\mathbf{x}} + \tilde{\boldsymbol{\varphi}}_x(\mathbf{x}, \hat{\mathbf{x}}, \mathbf{u})\,. \tag{9.8}$$

is obtained. For this system a standard HGO based on the approach in [BZH02] is not possible.

But from the results of Chap. 8 it is known that the feedback-linearized system in the state space of $\mathbf{z}$ fulfills all requirements for the HGO design. According to Eq. (8.30) it is given by

$$\dot{\mathbf{z}} = \mathbf{A}_z \cdot \mathbf{z} + \mathbf{B}_z \cdot (\mathbf{v} - \boldsymbol{\eta}) \tag{9.9a}$$

$$\mathbf{y} = \mathbf{C}_z \cdot \mathbf{z} \tag{9.9b}$$

and can also be expressed in the form of Eq. (9.5)

$$\dot{\mathbf{z}} = \mathbf{A}_z \cdot \mathbf{z} + \boldsymbol{\varphi}_z(\mathbf{z}, \mathbf{v}) \tag{9.10a}$$

$$\mathbf{y} = \mathbf{C}_z \cdot \mathbf{z} \tag{9.10b}$$

with $\mathbf{A}_z = \mathbf{A}_x$ and $\mathbf{C}_z = \mathbf{C}_x$. For this model an HGO design in the state space of $\mathbf{z}$ is possible.

But this direct HGO design would neglect one aspect. The linearizing feedback requires information about the entire state vector $\mathbf{z}$ and is consequently also affected by the observation (O). To further study this point and also keeping the influence of $\boldsymbol{\eta}$ in mind – which can be interpreted as disturbance at the system input (v) – the following system model is introduced instead of Eq. (9.10).

$$\dot{\mathbf{z}} = \mathbf{A}_z \cdot \mathbf{z} + \mathbf{B}_z \cdot \mathbf{v} + \mathbf{B}_z \cdot \boldsymbol{\delta}_{vO}(\mathbf{z}, \hat{\mathbf{z}}, \mathbf{v}) \qquad \text{(system)}, \tag{9.11}$$

where $\boldsymbol{\delta}_{vO}$ is interpreted as disturbance considering the above mentioned aspects. From Fig. 9.1 it is obvious that the stabilizing feedback affects the plant in the same way like the observer. Since the feedback matrix has been derived based on the same system model, which is also part of the observer – due to its high gains minimizing the influence of disturbances – a linear observer structure can be assumed in the following

$$\dot{\hat{\mathbf{z}}} = \mathbf{A}_z \cdot \hat{\mathbf{z}} + \mathbf{B}_z \cdot \mathbf{v} + \mathbf{H}_z \cdot (\mathbf{y} - \mathbf{C}_z \cdot \hat{\mathbf{z}}) \qquad \text{(observer)}. \tag{9.12}$$

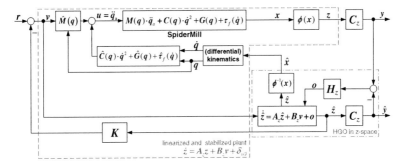

Figure 9.1: Feedback linearized system with HGO

As a consequence of the linearization and stabilization, the non-linear observer (9.7) is split into two linear ones. Each subsystem (9.13a) has its own observer (9.13b), whereas the linear parts of both subsystems are identical. Since also both observer-structures are equal, the same observer can be designed for both subsystems.

$$\dot{\mathbf{z}}^i = \mathbf{A}_{i,z} \cdot \mathbf{z}^i + \mathbf{B}_{i,z} \cdot v_i + \boldsymbol{\delta}_{vO,i}\left(\mathbf{z}, \hat{\mathbf{z}}, \mathbf{v}\right), \qquad i = 1, 2 \tag{9.13a}$$

$$\dot{\hat{\mathbf{z}}}^i = \mathbf{A}_{i,z} \cdot \hat{\mathbf{z}}^i + \mathbf{B}_{i,z} \cdot v_i + \mathbf{H}_{i,z} \cdot \left(y_i - \mathbf{C}_i \cdot \hat{\mathbf{z}}^i\right), \qquad i = 1, 2, \tag{9.13b}$$

with $\mathbf{H}_{i,z} = \begin{bmatrix} h_1^i & h_2^i \end{bmatrix}^T$, $\mathbf{z}^i = \begin{bmatrix} z_1^i & z_2^i \end{bmatrix}^T$, $\hat{\mathbf{z}}^i = \begin{bmatrix} \hat{z}_1^i & \hat{z}_2^i \end{bmatrix}^T$, $\boldsymbol{\delta}_{vO,i} = \begin{bmatrix} 0 & \delta_{vO,i} \end{bmatrix}^T$. The matrices $\mathbf{A}_{i,z}$, $\mathbf{B}_{i,z}$ and $\mathbf{C}_{i,z}$ are given in Eq. (8.22), guaranteeing the observability of $(\mathbf{A}_{i,z}\mathbf{C}_{i,z})$. In Eq. (9.13) a linear observer and a linear system with the disturbance $\boldsymbol{\delta}_{vO,i}$ are given. In the following the influence of $\boldsymbol{\delta}_{vO,i}$ on the linear observer is further analyzed. Thereto the estimation error is calculated

$$\dot{\tilde{\mathbf{z}}}^i = \left(\mathbf{A}_{i,z} - \mathbf{H}_{i,z} \cdot \mathbf{C}_{i,z}\right) \cdot \tilde{\mathbf{z}}^i + \boldsymbol{\delta}_{vO,i}\left(\mathbf{z}, \hat{\mathbf{z}}, \mathbf{v}\right). \tag{9.14}$$

It is obvious that $\tilde{\mathbf{z}}^i$ is influenced by the disturbance $\boldsymbol{\delta}_{vO,i}$. The disturbance is a consequence of the modeling errors and a wrong estimation of the system states. The influence of the modeling errors is reduced by the stabilizing state feedback, cp. Sec. 8.4. Applying it the nonlinear part $\boldsymbol{\delta}_{O,i}$ mainly represents the estimation error of the observer, which influences the linearizing feedback

$$\dot{\tilde{\mathbf{z}}}^i = \left(\mathbf{A}_{i,z} - \mathbf{H}_{i,z} \cdot \mathbf{C}_{i,z}\right) \cdot \tilde{\mathbf{z}}^i + \boldsymbol{\delta}_{O,i}\left(\mathbf{z}, \hat{\mathbf{z}}\right). \tag{9.15}$$

For further discussing the aspect of high observer gains in case of the SpiderMill the same analysis as in [Kha02] is performed. Thereto the matrices introduced in the context of Eq. (8.22) and (9.13) are inserted in Eq. (9.15) yielding

$$\dot{\tilde{z}}_1^i = \tilde{z}_2^i - h_1^i \cdot \tilde{z}_1^i \tag{9.16a}$$

$$\dot{\tilde{z}}_2^i = \delta_{v,i} - h_2^i \cdot \tilde{z}_1^i. \tag{9.16b}$$

and consequently the following transfer functions

$$G_1(s) = \frac{\tilde{z}_1^i}{\delta_{v,i}} = \frac{1}{s^2 + h_1^i \cdot s + h_2^i} \quad \text{and} \quad G_2(s) = \frac{\tilde{z}_2^i}{\delta_{v,i}} = \frac{s + h_1^i}{s^2 + h_1^i \cdot s + h_2^i}. \tag{9.17}$$

From Eq. (9.17) it is obvious that for extra high controller gains $h_2^i \gg h_1^i \gg 1$ the influence of the non-linear part vanishes and that it is consequently justified to assume the observer as linear. A general discussion of this aspect can be found in [DK97, Kha02, BZH02].

In case of the SpiderMill the consequences for the observer design are: the eigenvalues of $(\mathbf{A}_{i,z} - \mathbf{H}_{i,z} \cdot \mathbf{C}_{i,z})$ have to be stable and in order to not degrade the performance of the state-space control they should also be n times ($n = 2 \ldots 6$) [FPEN06] larger than that of the system stabilized by the state feedback.

This means linear control theory is applied in the following for the design of the observer [Ż03, DB05, FPEN06], strongly simplifying it. But since the feedback linearization is not exact and the gains of the stabilizing feedback and the observers are finite, the mutual interactions between the linearized system and the observers have to be analyzed while dimensioning the gains of both the observer and the stabilizing feedback.

9.1.2 Design of an HGO

Based on the considerations above two linear observers will be designed in this section for the linearized and stabilized system. The observers will be applied for estimating the states $\hat{\mathbf{z}}$ taking into account the different $\mathbf{K}_i$s (cp. Tab. 8.1).

Since the estimation error should converge much faster than the control error of the state-space control, the eigenvalues of the observer have to be chosen negative and smaller than that of the linearized and stabilized system. The chosen values and the belonging gains are summarized in Tab. 9.1.

For the verification of the observers a double integrator system with stabilizing feedback is used. The reference-input is set to $r_i = 0$ in order to study the eigenmotion of the system. For the test case the integrators of the reference plant and that of the observer are initialized with

$$z_{1,0}^i = 1, \quad z_{2,0}^i = 0 \quad \text{and} \quad \hat{z}_{1,0}^i = 0, \quad \hat{z}_{2,0}^i = 0.5. \tag{9.18}$$

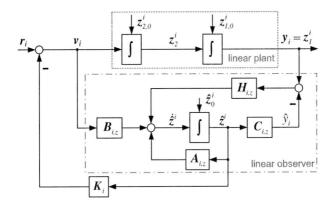

Figure 9.2: Linear observer and linear, stabilized subsystem

combination	observer		controller	
Comb. 1	$\lambda_{b1}^i = -690$ $h_1^i = 1740$	$\lambda_{b2}^i = -1050$ $h_2^i = 724500$	$\lambda_1^i = -230$ $k_{Pi} = 80500$	$\lambda_2^i = -350$ $k_{Di} = 580$
Comb. 2	$\lambda_{b1}^i = -180$ $h_1^i = 480$	$\lambda_{b2}^i = -300$ $h_2^i = 54000$	$\lambda_1^i = -60$ $k_{Pi} = 6000$	$\lambda_2^i = -100$ $k_{Di} = 160$

Table 9.1: Combinations of high gain observers and controllers

In case of both observer-controller combinations the stabilizing state feedback compensates the initial deflection (here: wrong initialization). Since in both cases the eigenvalues of the observers have been chosen three times smaller (on the left-hand side of the controller eigenvalues in the complex s-plane) than that of the controllers the observers converge faster. Due to the higher gains of Comb. 1 the convergence of both the observer and the eigenmotion is much faster than for Comb. 2 – cp. Fig. 9.3 and 9.4. But also the amplification of the wrong initial states is higher. For that reason a to some degree slower convergence of the observation error is in some situations more useful, which is obtained by Comb. 2. A further aspect, which cannot be seen at this step, is the fact that the high gains of Comb. 1 will lead to a bang-bang behavior of the control variable (cp. Sec. 10.3.2). This is the main reason, why Comb. 2 has been introduced after some tuning experiments. In general, a compromise between the transient behavior, the convergence of the observation and the course of the control variables has to be found.

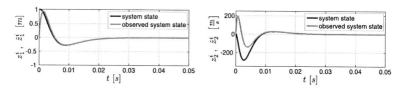

Figure 9.3: Course of system $\mathbf{z}^i$ and estimated system states $\hat{\mathbf{z}}^i$ (Comb. 1)

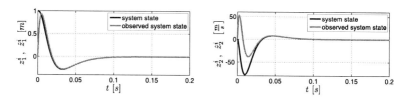

Figure 9.4: Course of system $\mathbf{z}^i$ and estimated system states $\hat{\mathbf{z}}^i$ (Comb. 2)

9.1.3 Implementation on the SimMechanics model

In this section the observers will be tested on the SimMechanics model of the robot. As reference system two ideal, decoupled double integrators are used, representing the exact linearized subsystems according to Eq. (8.22). Also for their stabilization a state feedback via $\mathbf{K}_i$ is implemented. The states of the reference system are measured and not observed. The described test environment is except for the observer equivalent to that in Sec. 8.4. Here it is used to evaluate the influence of the disturbance $\boldsymbol{\delta}_{vO}$ in case of the defined test scenario.

Comparing the simulation results shown in Fig. 9.5 for Comb. 1 with that for the same scenario without observer in Fig. 8.6 it becomes obvious that the courses of the errors are similar, as well as their scale. The spindle length error Δz_1^k in Fig. 9.5 is insignificantly larger, since it is additionally influenced by the observer. In contrast the velocity error Δz_2^k is marginally lower, which is a consequence of the fact, that both the observer and the reference system take a double integrator as plant. It can be concluded that the errors mainly result from the model inaccuracy.

Also Comb. 2 is tested for the scenario. From the comparison of the results in Fig 9.6 and 8.7 it can be seen, that again the errors are of the same scale and only marginally worse applying the observer.

Additionally the forces for both combinations are compared to each other in Fig. 9.7. From it the large difference of the control inputs for both combinations

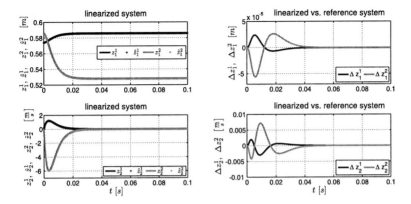

Figure 9.5: Linearized system with observed states ↔ reference system (Comb. 1)

becomes obvious. As already discussed in Sec. 9.1.2 only a fast convergence (transient behavior) of the control and observation errors is not the major goal of the here performed design. A compromise is necessary if effects like actuator limitation and wear shall be handled. For that reason Comb. 2 is also one of the applied strategies for the trajectory tracking control in Chap. 10.

9.2 The discrete Linear Kalman-Filter (LKF)

9.2.1 Theory of the LKF

Another approach to estimate the spindle velocities (and length) is to describe the system by discrete stochastic dynamical equations and to apply the discrete Linear Kalman-Filter (LKF). It is called linear since it is assumed, that the system and the measurements are governed by linear equations. The LKF operates by propagating the mean and covariance of the system state through time. The theory of the LKF, shortly summarized below is mainly a recapitulation from [Sim06, Neg03], where a detailed discussion can be found.

The discrete time dynamic system is given by

$$\mathbf{x}_k = \mathbf{A}_{k-1}\mathbf{x}_{k-1} + \mathbf{B}_{k-1}\mathbf{u}_{k-1} + \mathbf{w}_{k-1} \qquad (9.19a)$$

$$\mathbf{y}_k = \mathbf{C}_k\mathbf{x}_k + \mathbf{v}_k \qquad (9.19b)$$

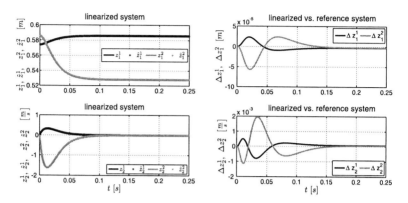

Figure 9.6: Linearized system with observed states $\leftrightarrow$ reference system (Comb. 2)

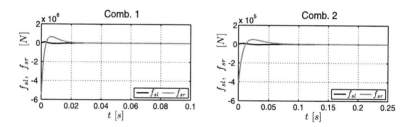

Figure 9.7: Spindle forces for Comb. 1 and 2

where $\mathbf{A}$ and $\mathbf{C}$ are the state transition and observation matrix, respectively. The noise processes $\mathbf{w}_k$ and $\mathbf{v}_k$ are white, zero-mean, uncorrelated, and have known covariances $\mathbf{Q}_k$ and $\mathbf{R}_k$:

$$\mathbf{w}_k \sim (\mathbf{0}, \mathbf{Q}_k), \quad \mathbf{v}_k \sim (\mathbf{0}, \mathbf{R}_k),$$

$$E[\mathbf{w}_k \mathbf{w}_j^T] = \mathbf{Q}_k \cdot \delta_{k-j}, \quad E[\mathbf{v}_k \mathbf{v}_j^T] = \mathbf{R}_k \cdot \delta_{k-j}, \quad E[\mathbf{v}_k \mathbf{w}_j^T] = \mathbf{0}, \tag{9.20}$$

where δ_{k-j} is the Kronecker delta function (for $k = j$: $\delta_{k-j} = 1$ elsewise $\delta_{k-j} = 0$). The goal is to estimate the state $\mathbf{x}_k$ based on the knowledge of system dynamics and the noisy measurements of $\mathbf{y}_k$. Two types of estimations have to be distinguished, depending on how many informations are available for the estimation of $\mathbf{x}_k$ [Sim06]:

$$\hat{\mathbf{x}}_k^+ = E[\mathbf{x}_k \mid \mathbf{y}_1, \mathbf{y}_2, \cdots, \mathbf{y}_k] \quad = \text{'a posteriori estimate'} \qquad (9.21a)$$
$$\hat{\mathbf{x}}_k^- = E[\mathbf{x}_k \mid \mathbf{y}_1, \mathbf{y}_2, \cdots, \mathbf{y}_{k-1}] = \text{'a priori estimate'} \qquad (9.21b)$$

where

$\hat{\mathbf{x}}_k^- = $ estimate of $\mathbf{x}_k$ before processing the measurement at time step k (9.22a)
$\hat{\mathbf{x}}_k^+ = $ estimate of $\mathbf{x}_k$ after processing the measurement at time step k. (9.22b)

The LKF-algorithm consists of three main steps [Neg03]:

Initialization: To initialize the LKF the initial posterior state estimate $\hat{\mathbf{x}}_0^+$ and the initial uncertainty of this state estimate $\mathbf{P}_0^+$ has to be specified[3]. Since $\hat{\mathbf{x}}_0^+$ is the initial estimate of $\mathbf{x}_0$ before any measurements are available, it is reasonable to chose $\hat{\mathbf{x}}_0^+$ as the expected value of the initial state $\mathbf{x}_0$:

$$\hat{\mathbf{x}}_0^+ = E(\mathbf{x}_0). \qquad (9.23)$$

If the initial state is perfectly known $\mathbf{P}_0^+ = 0$. If no information about it are available[4] $P_0^+ = \infty \mathbf{I}$. In general it is the uncertainty of the initial estimate of $\mathbf{x}_0$

$$\mathbf{P}_0^+ = E[(\mathbf{x}_0 - \hat{\mathbf{x}}_0^+)(\mathbf{x}_0 - \hat{\mathbf{x}}_0^+)^T]. \qquad (9.24)$$

Prediction: At each time step the system can be in a different system state. Hence, the LKF calculates a new a priori estimate for its state $\hat{\mathbf{x}}_k^-$ for each time step k. The prediction (also: time update or propagation) equations predetermine the new system state by projecting forward the most recent belief of it:

$$\hat{\mathbf{x}}_k^- = \mathbf{A}_{k-1}\hat{\mathbf{x}}_{k-1} + \mathbf{B}_{k-1}\mathbf{u}_{k-1} = \text{'a priori state estimate'} \qquad (9.25a)$$
$$\mathbf{P}_k^- = \mathbf{A}_{k-1}\mathbf{P}_{k-1}^+\mathbf{A}_{k-1}^T + \mathbf{Q}_{k-1}. \qquad (9.25b)$$

Correction: The correction (also: measurement update) equations integrate the measurements and can therefore only be applied if there is a measurement available. Since that gives a direct information about the current state, it can be used to correct the most recent belief. The equations calculate the a posterior estimate $\hat{\mathbf{x}}_k^+$ as a linear combination of an a priori state estimate $\hat{\mathbf{x}}_k^-$ and a weighted difference

---

[3]The term $\mathbf{P}_k$ is used to denote the covariance of the estimation error, where $\mathbf{P}_k^-$ denotes that of $\hat{\mathbf{x}}_k^-$ and $\mathbf{P}_k^+$ that of $\hat{\mathbf{x}}_k^+$.
[4]The identity matrix is given by $\mathbf{I}$ in order to distinguish it from the expectation value $E(\,\cdot\,)$.

between the actual measurement $\mathbf{y}_k$ and a measurement prediction $\mathbf{C}_k\hat{\mathbf{x}}_k^-$

$$\hat{\mathbf{x}}_k^+ = \hat{\mathbf{x}}_k^- + \mathbf{K}_k(\mathbf{y}_k - \mathbf{C}_k\hat{\mathbf{x}}_k^-) = \text{'a posteriori state estimate'} \qquad (9.26a)$$

$$\begin{aligned}
\mathbf{P}_k^+ &= (\mathbf{I} - \mathbf{K}_k\mathbf{C}_k)\mathbf{P}_k^-(\mathbf{I} - \mathbf{K}_k\mathbf{C}_k)^T + \mathbf{K}_k\mathbf{R}_k\mathbf{K}_k^T \\
&= [(\mathbf{P}_k^-)^{-1} + \mathbf{C}_k^T\mathbf{R}_k^{-1}\mathbf{C}_k]^{-1} \\
&= (\mathbf{I} - \mathbf{K}_k\mathbf{C}_k)\mathbf{P}_k^-.
\end{aligned} \qquad (9.26b)$$

The difference $\mathbf{y}_k - \mathbf{C}_k\hat{\mathbf{x}}_k^-$ in Eq. (9.26a) is called the *measurement innovation* or the *residual* [Neg03] and reflects the discrepancy between the predicted measurement $\mathbf{C}_k\hat{\mathbf{x}}_k^-$ and the actual measurement $\mathbf{y}_k$. The $n \times m$ matrix $\mathbf{K}_k$ is chosen to be the *gain* or *blending factor* that minimizes the a posteriori error covariance defined in Eq. (9.26b). It is the so-called *Kalman-Filter gain* and is given by

$$\begin{aligned}
\mathbf{K}_k &= \mathbf{P}_k^-\mathbf{C}_k^T(\mathbf{C}_k\mathbf{P}_k^-\mathbf{C}_k^T + \mathbf{R}_k)^{-1} \\
&= \mathbf{P}_k^+\mathbf{C}_k^T\mathbf{R}_k^{-1}
\end{aligned} \qquad (9.27)$$

The new a posterior state estimate is used in the next time step for calculating the new a priori state estimate. This makes the recursive nature of the LKF obvious. The fact, that not all data is required for the calculation allows its effective practical implementation.

Selected equations: From the alternative possible choices here the first expression for $\mathbf{K}_k$, cp. Eq. (9.27), is used because it just requires the a priori covariance and the third expression for $\mathbf{P}_k^+$, cp. Eq. (9.26b), due to its lower computational burden.[5]

Algorithm: *The calculation steps for the LKF can be summarized to [Sim06]:*

(1) The dynamic system is given by:

$$\mathbf{x}_k = \mathbf{A}_{k-1} \cdot \mathbf{x}_{k-1} + \mathbf{B}_{k-1} \cdot \mathbf{u}_{k-1} + \mathbf{w}_{k-1}$$

$$\mathbf{y}_k = \mathbf{C}_k \cdot \mathbf{x}_k + \mathbf{v}_k$$

$$E[\mathbf{w}_k\mathbf{w}_j^T] = \mathbf{Q}_k \cdot \delta_{k-j}, \quad E[\mathbf{v}_k\mathbf{v}_j^T] = \mathbf{R}_k \cdot \delta_{k-j}, \quad E[\mathbf{v}_k\mathbf{w}_j^T] = \mathbf{0}. \qquad (9.28)$$

(2) The LKF is initialized to:

$$\hat{\mathbf{x}}_0^+ = E(\mathbf{x}_0) \qquad and \qquad \mathbf{P}_0^+ = E[(\mathbf{x}_0 - \hat{\mathbf{x}}_0^+)(\mathbf{x}_0 - \hat{\mathbf{x}}_0^+)^T] \qquad (9.29)$$

---

[5]The first equation for the calculation of $\mathbf{P}_k^+$ known as Joseph's form is more stable and robust than the here used third expression.

(3) The LKF is given by the following equations, which have to be calculated for each time step k = 1, 2,...:

$$\mathbf{P}_k^- = \mathbf{A}_{k-1}\mathbf{P}_{k-1}^+\mathbf{A}_{k-1}^T + \mathbf{Q}_{k-1}$$
$$\mathbf{K}_k = \mathbf{P}_k^-\mathbf{C}_k^T(\mathbf{C}_k\mathbf{P}_k^-\mathbf{C}_k^T + \mathbf{R}_k)^{-1}$$
$$\hat{\mathbf{x}}_k^- = \mathbf{A}_{k-1}\hat{\mathbf{x}}_{k-1} + \mathbf{B}_{k-1}\mathbf{u}_{k-1} = \text{'a priori state estimate'}$$
$$\hat{\mathbf{x}}_k^+ = \hat{\mathbf{x}}_k^- + \mathbf{K}_k(\mathbf{y}_k - \mathbf{C}_k\hat{\mathbf{x}}_k^-) = \text{'a posteriori state estimate'}$$
$$\mathbf{P}_k^+ = (\mathbf{I} - \mathbf{K}_k\mathbf{C}_k)\mathbf{P}_k^- \tag{9.30}$$

9.2.2 Filter parameters

Since only the spindle length are determined in case of the SpiderMill in the following the scalar case is discussed. Usually the measurement noise covariance R_k is determined before the LKF is applied. It is always possible to take some sample measurements, since its has to be possible to measure the process in order to be able to use the LKF. For that purpose the motor encoders are used. From Eq. (9.19b) it follows, that the measurement of the output y_k is equal to the latest encoder count times the resolution $\theta_m^\star$ ('true' position at $k \cdot T_S$) with an error v_k. This error is here considered to be only due to the quantization, neglecting other influences. It is assumed, that it is white, bounded, zero mean and uniformly distributed such that $-\theta_m^\star \leq v_k \leq \theta_m^\star$, cp. [JSHC00]. Based on this considerations, the measurement error covariance R can be calculated. First, the probability density function of v_k is determined to

$$f(v_k) = \frac{1}{2\theta_m^\star}. \tag{9.31}$$

Resulting in an expectation value $E(v_k)$ of

$$\mathrm{E}(v_k) = \int_{-\theta_m^\star}^{\theta_m^\star} v_k \cdot f(v_k)\, \mathrm{d}v_k = \int_{-\theta_m^\star}^{\theta_m^\star} v_k \cdot \frac{1}{2\theta_m^\star}\, \mathrm{d}v_k = 0 \tag{9.32}$$

and a variance of

$$\sigma^2(v_k) = \mathrm{E}(v_k^2) - (\mathrm{E}(v_k))^2 = \frac{\theta_m^{\star 2}}{3}. \tag{9.33}$$

With Eq. (9.33) the covariance $\mathrm{COV}(v_k)$ i.e. R of quantization noise v_k is given, since $\sigma^2(v_k) = \mathrm{COV}(v_k)$ holds.

Generally, the determination of the process noise covariance Q is more difficult than that of R, since it is normally not possible to observe the process which is

estimated [WB06]. But also in cases without a rational basis for the choice of
the filter parameters a good filter performance [WB06] can be achieved by their
tuning. This tuning is usually performed offline by simulating the system model.
Another, in no case perfect, but here applied way for choosing Q is to compute the
disturbance $\boldsymbol{\eta}$ of the feedback-linearized model of the SpiderMill with Eq. (8.28).
In this context, it is assumed that there is an estimation error in case of the parameter identification and it is further assumed, that the parameters are constant and
not changing during the process. But in case of a constant error, the disturbance
cannot be considered as Gaussian white noise. This aspect is analyzed below for
different percentages of parameter uncertainty of the mass an inertia parameters of
the SpiderMill. With the uncertain parameters $\hat{m}_i$ and $\hat{I}_{mi}$ the uncertain matrices
$\hat{\mathbf{M}}$, $\hat{\mathbf{C}}$, $\hat{\mathbf{G}}$ and $\hat{\mathbf{F}}$ are determined, whereas for the calculation of $\mathbf{M}$, $\mathbf{C}$, $\mathbf{G}$, $\mathbf{F}$ the
estimated parameters are used. Applying Eq. (8.28) $\boldsymbol{\eta}$ is calculated for the differ-
ent trajectories planned in Sec. 7.3. Exemplarily the results for the trajectories
without a jerk limit $|\dot{\boldsymbol{\tau}}| \to \infty$ in case of a 8% parameter uncertainty are shown in
Fig. 9.8. For lower jerk limits also the $\boldsymbol{\eta}_k$s are lower.

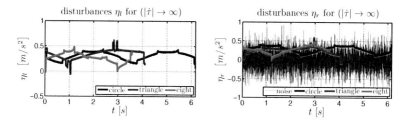

Figure 9.8: Disturbances $\boldsymbol{\eta}_k$ for unlimited jerk

Even though the disturbance Q is not Gaussian white noise, the limits of the disturbance are useful information for the LKF design. From Fig. 9.8 it can be seen
that the most trajectories (especially that with a limited jerk) have a maximal
disturbance of 0.05 m/s². For symmetry reasons −0.05 m/s² and 0.05 m/s² are
assumed as lower and upper bounds. Based on these bound a proper covariance Q
for the process noise is defined. For that purpose it is assumed, that the process
noise is pure Gaussian white noise. For a covariance $Q = 0.06$ the white Gaussian
noise, shown in Fig. 9.8 (right), has almost the same bounds as defined above. For
hat reason the covariance of the process noise w_k is assumed to $Q = 0.06$ in case
of an 8% parameter uncertainty.

percentage [%]	Q	$\overline{\Delta P}$ $[m]$	$\overline{\Delta S}$ $[m/s]$	$\overline{\Delta \dot{S}}$ $[m/s^2]$	$\overline{\eta}$
4	$7 \cdot 10^{-3}$	$2.0290 \cdot 10^{-4}$	$3.8315 \cdot 10^{-5}$	$4.4015 \cdot 10^{-3}$	99.9793
8	$6 \cdot 10^{-2}$	$1.7741 \cdot 10^{-4}$	$3.3473 \cdot 10^{-5}$	$4.3241 \cdot 10^{-3}$	99.9796
30	1	$1.7834 \cdot 10^{-4}$	$3.3270 \cdot 10^{-5}$	$4.3481 \cdot 10^{-3}$	99.9796

Table 9.2: Percentages of parameter uncertainty and belonging Q

In total three percentages of parameter uncertainty have been studied and sum-marized in Tab. 9.2[6]. Based on the results of Tab. 9.2 – including a lookahead of the trajectory tracking performance of the three different resulting LKFs – the covariance is chosen to $Q = 0.06$ for the design of the LKF in the following.

For the case where Q and R are constant, both $\mathbf{P}_k$ and $\mathbf{K}_k$ converge quickly and remain constant, cp. Eq. (9.30). This phenomenon is known as 'fall asleep'. For this case the filter parameters can be pre-computed offline. But more frequently Q does not remain constant. For example if unknown disturbances act on the Spi-derMill or if it is used for milling applications the process noise Q has to become Q_k in order to better consider these effects. Since in this thesis only a trajectory tracking (position control) and no hybrid control is studied Q and R are assumed as constant.

9.2.3 Some remarks on the LKF

The LKF has the same structure as an observer (Luenberger), with the differ-ence that it is not designed via pole-placement, but based on an optimal control approach. As in case of the HGO the separation principle is valid and the ap-proximated value of the state $\hat{x}$ can be used for the control. The use of the LKF suggests also to design the state feedback in form of an optimal control instead of using pole-placement [Lun02]. In cases where an optimal controller is realized with the help of a LKF the control structure is called LQG-controller. This concept is amongst others applied in Sec. 10.4.1.

9.2.4 LKF in the test scenario

For a first verification of the LKF the same test scenario as for the HGOs, cp. Sec. 9.1.3, is applied using the controller gains of C2. From Fig. 9.9 it can be seen, that the LKF needs some steps for 'swinging in', where also the velocity error is

---

[6]The last four columns are performance indices defined in Sec. 10.2. Its values have been obtained by performing a realistic simulation with the controller gains of C2.

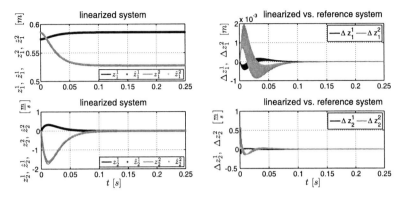

Figure 9.9: Test scenario for LKF (C2)

larger than in case of the HGOs. But the results of Chap. 10 show its superior performance in case of trajectory tracking control after the first steps.

9.3 Adaptive Windowing (AW)

A further approach for the velocity estimation is the Adaptive Windowing (AW) technique. In contrast to HGO and LKF, where the position and velocity have been estimated in case of the AW only the velocities are determined based on the encoder-signals of the motors. Both the HGO and the LKF base on a linear(ized) model of the plant. The observation results are correlated with the quality of the applied system model. Furthermore, both methods share the fundamental trade-offs between: noise reduction, control delay, estimation reliability and accuracy, error convergence rate and computational load [JSHC00]. Furthermore they always require a more or less long tuning process. The AW strategies below and their theoretical background are mainly taken from [JSHC00]. Since both spindle velocities are calculated separately the strategy is presented using scalar variables.

9.3.1 Non-model-based velocity determination

In cases where no adequate model of the physical system is available or the system is nonlinear and the velocity cannot be measured directly, the only way for determining it, is the usage of the discrete and quantized position signal from an encoder.

Again the output equation of the model (9.19b) is valid. It is assumed, that x_k is the true position (spindle length) at $k \cdot T_S$ and $y_k = x_k + v_k$ is its measurement. Where v_k denotes the measurement error, due to the quantized encoder signal. The error is assumed as zero mean, bounded, uniformly distributed. Consequently, $-\theta_m^\star \leq v_k \leq \theta_m^\star$ holds. The variance of the error is calculated by Eq. (9.33).

Beside the below introduced AW several methods based on the discrete and quantized position signal exists. A short discussion about their field of application can be found in [HJSC97, JSHC00]. Possible strategies are e.g. the Finite Difference Method (FDM, unsuitable for low velocities), the Inverse Time Approach (ITA, unsuitable for high velocities) and the Fixed-Filter-Methods (FFMs). The latter ones have the disadvantages, that their transfer functions become part of the closed loops and the required tuning process. Furthermore, the quality of the velocity estimation is very poor in case of fast velocity changes [JSHC00]. To avoid the above mentioned and further disadvantages AW strategies are commonly employed.

9.3.2 First-Order Adaptive Windowing algorithms

The theory of the AW algorithms summarized below is mainly taken from [JSHC00, HJSC97].

Motiviation: The First-Order Adaptive Windowing (FOAW) is optimal in the sense that it minimizes the velocity error variance while maximizing the estimation accuracy (requiring no tradeoff) [JSHC00]. All AW algorithms pose noise filtering properties, but preserve at the same time the velocity transients, in contrast to the FFMs. Their design requires the selection of only one parameter, a bound on the measurement noise, here selected according to Eq. (9.33).

The goal in case of FOAW is to find an estimate $\hat{z}_k$ for the velocity $z = \mathrm{d}x/\mathrm{d}t$ from the measurements $\{y_i\}_{k-n}^k$, with the windows size n. Furthermore, the algorithm should be used online (low computational burden) and should minimize effects of noise. Also the delays should be minimized in order to avoid a reduction of the phase margin of the closed-loop control. These goals are not reachable with the above discussed approaches using the discrete encoder signal only. In case of an effective estimation technique, high sampling rates and thereto higher controller gains – leading to lower tracking errors – should be possible. E.g. if HGOs are used the measurement noise – leading to an overshooting of the observer (cp. Sec. 9.1.3) – is strongly limiting the possible controller gains (at least theoretically).

Common aspects: The different FOAW algorithms base upon the Euler approximation [SSdc71]

$$\hat{z}_k = \frac{y_k - y_{k-1}}{T_S} = \frac{x_k - x_{k-1}}{T_S} + \frac{v_k - v_{k-1}}{T_S}, \tag{9.34}$$

which is more precise if the two position samples to which it is applied are far apart. This implies, that the variance of the velocity decreases with increasing window length n.

In its basic version, the end-fit first-order adaptive windowing (end-fit-FOAW), the last n velocity estimates obtained by Eq. (9.34) are averaged

$$\hat{\mathbf{z}}_k = \frac{1}{n} \sum_{j=0}^{n-1} \hat{\mathbf{z}}_{k-j} = \frac{\mathbf{y}_k - \mathbf{y}_{k-n}}{n\,T_S} \quad (= b_n : \text{'end-fit-FOAW'}). \tag{9.35}$$

For well sampled signals, an increasing of n is equivalent to a decreasing of T_S. According to [JSHC00] a large window introduces a time delay and reduces the estimation reliability. For a trade-off between 'precision' and 'reliability' n is selected adaptively depending on the signal itself in case of FOAW. For high velocities n has to be small (more reliable estimate; faster calculation) for low velocities large (more precise estimates). Noise reduction and precision define a lower and the reliability an upper bound for n. The key is to determine an appropriate n. The precision of the velocity determination increases with n. In order to find the largest possible n satisfying the accuracy requirements a criterion for its selection is defined in [JSHC00]. Its allows to decide, whether the slope of a straight line approximates reliably the derivative of the position signal between two samples x_k, x_{k-n} or not. It has to be ensured, that a straight line passing trough y_k, y_{k-n} is covering all intermediate samples within an uncertainty band defined by the peak norm of the noise [JSHC00]:

$$e = \|v_k\|_\infty, \quad \forall k. \tag{9.36}$$

The basic algorithm, end-fit-FOAW, estimates the velocity based on two position measurements. As a consequence undesired overshoots can occur for small n. To provide additional smoothing, normally, the best-fit-FOAW using all the samples inside the window is applied. In this case the velocity estimate is the slope of a least-square approximation (minimizing the error energy) [JSHC00]. In case of the best-fit-FOAW only the calculation of b_n changes.

Algorithms: *In case of the end-fit-FOAW and the best-fit-FOAW the 'optimal' windows length n is found by: $n = max\{1, 2, \ldots\}$ such that [JSHC00]:*

$$\left| y_{k-i} - y_{k-i}^L \right| \le e, \quad \forall i \in \{1, 2, \ldots, n\} \tag{9.37}$$

where

$$y^L_{k-i} = a_n + b_n(k-i)T_S \quad \text{('straight line'), with:} \tag{9.38a}$$

$$a_n = \frac{ky_{k-n} + (n-k)y_k}{n}, \quad \text{and} \tag{9.38b}$$

$$b_n = \hat{z}_k = \frac{1}{n}\sum_{i=0}^{n-1}\hat{\mathbf{z}}_{k-i} = \frac{\mathbf{y}_k - \mathbf{y}_{k-n}}{n\,T_S} \quad \text{'end-fit-FOAW'} \tag{9.38c}$$

(where b_n is the slope of the straight line passing through y_k and y_{k-i})

$$b_n = \hat{z}_k = \frac{n\sum_{i=0}^{n}y_{k-i} - 2\sum_{i=0}^{n}i\,y_{k-i}}{T_S\,n(n+1)(n+2)/6} \quad \text{'best-fit-FOAW'.} \tag{9.38d}$$

The optimality of the approach is justified by Prop. 2.

Proposition 2 *For a position trajectory with a piecewise continuous and bounded derivative – in case of a uniformly distributed measurement noise – the adaptive windowing algorithm minimizes the velocity error variance and maximizes the accuracy of the velocity estimate. Prop. 2 and its proof can be found in [JSHC00].*

The end-fit- and the best-fit-FOAW algorithms have the following steps [JSHC00]:

(1) Set index $i = 1$.

(2) Set y_k as the last measured sample and y_{k-i} as the ith before.

(3) Calculate b_n with Eq. (9.38c) or (9.38d).

(4) Check whether the straight line (9.38a) passes through all points inside the window within the defined uncertainty band (9.36).

(5) If true, set index $i = i + 1$ and goto (3). Else return last estimate.

Also the best-fit-FOAW is only suboptimal and the quality of the velocity estimation can be further improved by e.g. using the best-fit-FOAW-R or other approaches which are additionally more robust against outliers - cp. [JSHC00].

9.3.3 FOAW algorithms in the test scenario

For the implementation of the FOAW algorithms, its peak norm of the noise $e = \|v_k\|_\infty$ is chosen to $\theta^\star_m$. To get a first impression of the performance of the algorithms the standard test scenario is applied again. Like in Sec. 9.2.4 the algorithms are used in combination with the feedback-linearized system and the PD controller (C2). In contrast to HGO and LKF the position signal is used directly. The results for the end-fit-FOAW are given in Fig. 9.10, that of the best-fit-FOAW in Fig. 9.11. The windows lengths in both cases are summarized in Fig. 9.12.

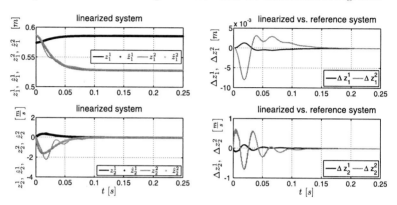

Figure 9.10: Test scenario for end-fit-FOAW (C2)

The quality of the estimation is for the best-fit-FOAW slightly better than for the end-fit-FOAW. The results of the first one are comparable to that of the LKF also with respect to its speed of convergence. Effects of a higher computational burden in case of the best-fit-FOAW could not be observed in the simulations and the practical implementation (cp. Chap. 10).

9.4 Conclusions and future work

In this chapter two different classes of strategies for determining the spindle-velocities of the SpiderMill have been introduced. The strategies of the first class are all model-based and realize some kind of estimation. But these strategies, namely the HGO and the LKF, can only be applied to linear or like here linearized systems. A possible extension in order to allow an state estimation in case of a non-linear dynamic model can be the usage of the Extended- [BD67,Sim06] or the Unscented-Kalman-Filter [JK97,JU04]. Whereas especially in the latter case the computational burden should not be underestimated in case of (parallel) robots. The other class of strategies, the AW algorithms have the main advantage, that

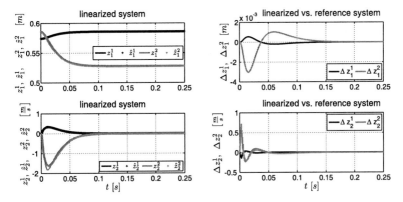

Figure 9.11: Test scenario for best-fit-FOAW (C2)

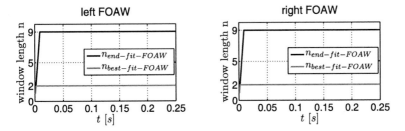

Figure 9.12: Windows lengths in case of the test scenario (C2)

they only require a precise position measurement and some statistical information about the noise. A dynamic model is not required. Whereas a model improvement, adaption or at least a change of its parameter in the first case requires a redesign of the observer or at least a new tuning, such steps are not necessary in case of AW. Beside the two algorithms introduced above further versions are available which perhaps further improve the velocity determination.

The performance of the introduced strategies will be further evaluated in case of the trajectory tracking control in Chap. 10. In this context only the estimated velocity signals will be used, whereas in case of all algorithms the position signals are taken directly from the encoders.

10 Trajectory tracking control

The most common task in robotics is the tracking of pre-defined trajectories. In order to follow them with an as small as possible trajectory tracking error sophisticated control strategies are required. In this chapter several commonly known model-based and non-model-based control algorithms are implemented for the SpiderMill and their performances are evaluated for the reference trajectories.

The chapters before set the stage for the design and implementation of the control strategies. Their interrelationship is shortly discussed before the controllers are designed: In Chap. 5 full- and reduced dynamic models of the SpiderMill have been derived. One of the first class, whose parameters have been identified in Chap. 6 is applied for the design of the model based-control strategies. The other ones are used for the development of non-model-based concepts. Several reference trajectories have been designed in Chap. 7 based on a new trajectory planning strategy in order to evaluate the performance of the control concepts derived in this chapter. The most of the strategies, especially the sophisticated ones, thereby require a knowledge of the spindle-velocities. In order to avoid their calculation by numerical derivation, several advanced strategies have been analyzed in Chap. 9.

In this chapter all parts come together for developing different trajectory tracking control strategies for the SpiderMill. Thereby, one of the concepts is analyzed in more detail, in order to understand the particularities in case of the SpiderMill better. For the development of the reference control concept the feedback-linearized and stabilized system derived in Sec. 8 in combination with the velocity reconstruction strategies, is extended by a tracking controller with linear and asymptotic converging error dynamics. The concept, known in literature as *joint space inverse dynamics control* has been chosen, since the influence of uncertainties on it has been discussed above in detail. All of the control concepts applied in this chapter are commonly known standard approaches. Their theoretical background can be found amongst others in [SS05, Cra05, SHV06].

Depending on the used control hardware, some of the here presented model-based control concepts cannot be directly applied for the real-time control of parallel robots with redundancies. The reason is the high computational load of the dynamic models obtained without performing a model simplification or reduction. In this thesis a real-time calculable model has been derived by applying the proposed simplification strategy. It defines the basis for all presented model-based concepts.

This chapter is structured as follows: In Sec. 10.1 similarities and differences between the simulation and the practical realization of the tracking control are discussed. Performance indices allowing a comparison of all developed strategies are defined in Sec. 10.2. In Sec. 10.3 a reference tracking controller is introduced and its design and performance are discussed. Further model-based control concepts are analyzed in Sec. 10.4 and non-model-based in Sec. 10.5. The simulation and measurement results are summarized in Sec. 10.6. Conclusions are drawn in Sec. 10.7.

10.1 Simulation and experimental setup

Before the control algorithms are introduced, the similarities and differences between the simulation environment and the experimental setup are shortly discussed below. In both cases the planned trajectories are realized as lookup tables. The controllers as well as the algorithms for the reconstruction of the spindle-velocities are implemented in an identical form.

The simulation environment in Matlab/Simulink has been described in Sec. 2.2.5. In order to simulate the MBS model like a physical plant a mixed simulation mode has been realized: A continuous integration algorithm is chosen for the overall simulation (simulating the MBS model with a variable step-size), while all other blocks, especially that of the controller and of the velocity reconstruction are simulated with a fixed sampling rate (here that of the dSPACE-system: $T_a = 0.001$ s). This mode has been implemented in order to allow a better comparison of the results, since in case of the practical implementation only a discrete solver with a fixed step-size can be chosen.

In this thesis two of the various possible simulation modes using the implemented simulation environment are used:

- Ideal simulation: The sensor and actuator models are turned-off.

- Realistic simulation: The sensor and actuator models are turned-on.

Furthermore the joints and the drive systems of the SimMechanics model are frictionless so far. This has a further consequence for some of the control algorithms. The algorithms compensating the influence of friction – here exemplarily all algorithms which use the feedback linearization described in Sec. 8.2 as part of the control concept – are realized in two different ways. For the simulation the friction compensation is turned off and enabled for the practical measurements.

For all of the following control algorithms the simulation and experimental results, applying the performance criteria defined in Sec. 10.2, are summarized in compact form in Sec. 10.6.

10.2 Performance criteria

In order to be able to compare the different control concepts in a compact form the following performance criteria are defined. They are equal to those in [Pie03], but here only time-independent ones are applied. The average trajectory error ΔP_i of trajectory i is given by

$$\Delta P_i = \frac{1}{N} \sum_{k=1}^{N} \sqrt{(x_{TCP,d}(t_k) - x_{TCP}(t_k))^2 + (y_{TCP,d}(t_k) - y_{TCP}(t_k))^2} \qquad (10.1)$$

where d stands for the desired values, t_k for the discrete sampling point and N for the number of sampling points. The mean absolute spindle error ΔS_i defines the error of the spindle coordinates in case of the ith trajectory

$$\Delta S_i = \frac{1}{2N} \left(\sum_{k=1}^{N} |(s_{l,d}(t_k) - s_l(t_k))| + \sum_{k=1}^{N} |(s_{r,d}(t_k) - s_r(t_k))| \right). \qquad (10.2)$$

An analogous index is defined for the spindle velocities

$$\Delta \dot{S}_i = \frac{1}{2N} \left(\sum_{k=1}^{N} |(\dot{s}_{l,d}(t_k) - \dot{s}_l(t_k))| + \sum_{k=1}^{N} |(\dot{s}_{r,d}(t_k) - \dot{s}_r(t_k))| \right). \qquad (10.3)$$

In order determine how close a planned trajectory is on the time-optimal one and to give an indicator for a bang-bang behavior of the drive torques the drive degree of utilization η_i is defined for the ith trajectory to

$$\eta_i = \left(1 - \frac{1}{N} \sum_{k=1}^{N} \min_{1 \le i \le n_{dr}} \left| \frac{|\tau_i(t_k)| - |\tau_i^{max}|}{\tau_i^{max}} \right| \right) \cdot 100\%, \qquad (10.4)$$

where n_{dr} is the number of drives and τ_i^{max} the maximum torque of drive i. For having criteria considering all of the different trajectories in case of one control concept together, the following mean values are introduced additionally

$$\overline{\Delta P} = \frac{1}{n_T} \sum_{i=1}^{n_T} \Delta P_i, \quad \overline{\Delta S} = \frac{1}{n_T} \sum_{i=1}^{n_T} \Delta S_i, \quad \overline{\Delta \dot{S}} = \frac{1}{n_T} \sum_{i=1}^{n_T} \Delta \dot{S}_i, \quad \overline{\eta} = \frac{1}{n_T} \sum_{i=1}^{n_T} \eta_i. \qquad (10.5)$$

where n_T is the number of the trajectories.

10.3 Reference tracking controller

10.3.1 Design of the reference tracking controller

In this section a reference trajectory tracking controller with linear and asymptotic converging error dynamics – known as *joint space inverse dynamics control* – is designed for the SpiderMill. Since the velocity estimation has only a marginal impact on the tracking error (cp. Sec. 9.1) it is neglected for the controller design. Again two identical controllers are designed for the two (coupled) subsystems starting using the model with uncertainty (8.32).

Based on the model (8.32) a controller leading to an asymptotic convergence of the dynamics of the trajectory tracking error has to be designed. For that purpose the reference input r_i is chosen to [SL91, SS05]

$$r_i = \ddot{q}_{ai,d} + k_{Di} \cdot \dot{q}_{ai,d} + k_{Pi} \cdot q_{ai,d}. \tag{10.6}$$

The input r_i is a function of the trajectory $q_{ai,d}$ of the ith active joints and its time-derivatives. With Eq. (5.27) and (8.19) the states can be determined from the active coordinates (which are here the spindle length)

$$z_{1,d}^i = q_{ai,d} \quad \text{and} \quad z_{2,d}^i = \dot{q}_{ai,d} \quad \text{with} \quad z_{2,d}^i = \dot{z}_{1,d}^i. \tag{10.7}$$

Using it the reference input r_i can be expressed by the states of system (8.32)

$$r_i = \dot{z}_{2,d}^i + k_{Di} \cdot z_{2,d}^i + k_{Pi} \cdot z_{1,d}^i. \tag{10.8}$$

Applying $\dot{z}_{1,d}^i = z_{2,d}^i$ the following state-space model can be derived from Eq. (10.8)

$$\begin{bmatrix} \dot{z}_{1,d}^i \\ \dot{z}_{2,d}^i \end{bmatrix} = \begin{bmatrix} 0 & 1 \\ -k_{Pi} & -k_{Di} \end{bmatrix} \cdot \begin{bmatrix} z_{1,d}^i \\ z_{2,d}^i \end{bmatrix} + \begin{bmatrix} 0 \\ 1 \end{bmatrix} \cdot r_i. \tag{10.9}$$

With the input (10.6) an asymptotic convergence of the trajectory tracking error shall be ensured. For analyzing this aspect Eq. (8.32) is subtracted from Eq. (10.9). The result is the state-space model of the tracking error

$$\begin{bmatrix} \dot{\tilde{z}}_1^i \\ \dot{\tilde{z}}_2^i \end{bmatrix} = \begin{bmatrix} 0 & 1 \\ -k_{Pi} & -k_{Di} \end{bmatrix} \cdot \begin{bmatrix} \tilde{z}_1^i \\ \tilde{z}_2^i \end{bmatrix} + \begin{bmatrix} 0 \\ 1 \end{bmatrix} \cdot \eta_i \tag{10.10}$$

with $(\tilde{\cdot})^i = (\cdot)_d^i - (\cdot)^i$. With the substitution

$$e_i = \tilde{z}_1^i \qquad \text{and} \qquad \dot{e}_i = \tilde{z}_2^i, \tag{10.11}$$

the following second order error dynamics is obtained from Eq. (10.10)

$$\ddot{e}_i + k_{Di} \cdot \dot{e}_i + k_{Pi} \cdot e_i = \eta_i. \tag{10.12}$$

The dynamics of the tracking error is analyzed in two steps. First the disturbance is neglected ($\eta_i = 0$). For this case the state-space model of the tracking error and consequently its dynamics is equal to that of the linearized and stabilized system (8.32), if $r_i = 0$ holds. This implies that the same gains as in Tab. 8.1 can be chosen in order to achieve the convergence of the trajectory tracking error.

In a second step the influence of the uncertainty η_i is analyzed. It is interpreted as disturbance, whose influence should be minimized. Comparing Eq. (10.10) and Eq. (8.32) it becomes obvious that the models are equal except for the sign of η_i (if $r_i = 0$ holds). For that reason the same procedure as in Sec. 8.4 is applied for analyzing its influence. The following transfer functions are obtained

$$G_{z1i}(s) = \frac{\tilde{z}_1^i}{\eta_i} = \frac{1}{s^2 + k_{Di} \cdot s + k_{Pi}} \tag{10.13a}$$

$$G_{z2i}(s) = \frac{\tilde{z}_2^i}{\eta_i} = \frac{s}{s^2 + k_{Di} \cdot s + k_{Pi}}. \tag{10.13b}$$

A comparison with Eq. (10.11) shows, that Eq. (10.13a) is equal to the position and Eq. (10.13b) to the velocity error. For verifying the influence of $\tilde{z}_1^i$ and $\tilde{z}_2^i$ on the tracking error the stationary final values of the position ($\eta_i = \frac{1}{s}$) and the velocity ($\eta_i = \frac{1}{s^2}$) are determined

$$\lim_{t \to \infty} y_i(t) = \lim_{s \to 0} s \cdot G_{z1i}(s) \cdot \eta_i = \frac{1}{k_{Pi}} \tag{10.14a}$$

$$\lim_{t \to \infty} y_i(t) = \lim_{s \to 0} s \cdot G_{z2i}(s) \cdot \eta_i = \frac{1}{k_{Pi}}. \tag{10.14b}$$

It is obvious that in case of step/ramp-like disturbances their influence in the stationary case depends on the gain k_{Pi}. Again the influence of the uncertainty η_i on the error dynamics (10.10) decreases with increasing k_{Pi}. Choosing a k_{Pi} according to Tab. 9.1 the influence of η_i on the tracking control will be marginal.

The analysis above and in Chap. 8 have been performed in order to study the individual parts of the joint space inverse dynamics control in the presence of modeling errors – the linearization and decoupling, the stabilization and the tracking control – separately. For closing its description the dynamics of the trajectory tracking error is given in terms of the active coordinates applying the control concept (without uncertainty)

$$\ddot{\tilde{\mathbf{q}}}_a + \mathbf{K}_D \cdot \dot{\tilde{\mathbf{q}}}_a + \mathbf{K}_P \cdot \tilde{\mathbf{q}}_a = \mathbf{0}. \tag{10.15}$$

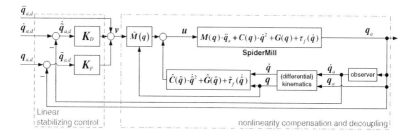

Figure 10.1: Reference tracking control

In contrast to *computed torque control* the *inverse dynamics control* requires an online calculation of the matrices of the feedback-linearization part. For that reason the strategy can in the most cases – and especially in case of complex parallel manipulators – not directly be applied to a mechanism. One way to possibly implement it despite its high computational load is to calculate and compensate only the dominant terms of the dynamic model [SS05]. But this directly lowers the performance of the tracking control. By contrast, based on the proposed model simplification strategy a compact model for straightforward applying the concept is obtained.

10.3.2 Simulation of the reference tracking control

In the following the reference tracking controller designed in Sec. 10.3.1 is verified using the simulation environment. The overall structure is given in Fig. 10.1. In a first step the two controller-observer combinations introduced in Sec. 9.1.2 (cp. Table 9.1) are analyzed in the context of trajectory tracking control. The eight trajectory is exemplarily chosen as reference input.

Realistic and ideal simulation - differentiating effect of the HGO

Realistic simulation: The results for performing a realistic simulation, while applying the two controller-observer combinations, are given in Fig. 10.2 to 10.4. From Fig. 10.2 it becomes obvious that the high gains of Comb. 1 lead to a bang-bang behavior of the spindle-forces, which can be avoided in case of Comb. 2. In contrast to the assumptions of Sec. 9.1.2 from Fig. 10.3 it can be seen, that the observation errors of the spindle lengths and velocities are even in the simulation smaller in case of Comb. 2. This does also hold for the tracking errors in Fig. 10.4.

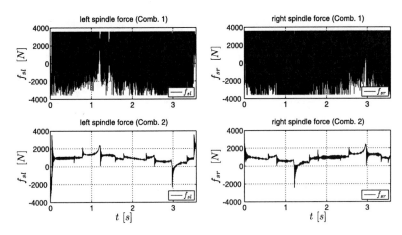

Figure 10.2: Spindle forces for Comb. 1 and 2 (realistic simulation)

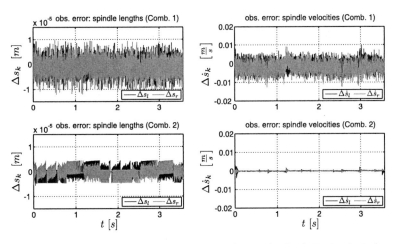

Figure 10.3: Observation errors in case of Comb. 1 and 2 (realistic simulation)

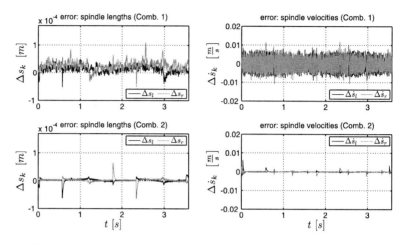

Figure 10.4: Tracking errors in case of Comb. 1 and 2 (realistic simulation)

The reason for the 'wrong' assumptions in Sec. 9.1.2 is the fact that there a set point control has been studied, where high gains lead to a better performance. In case of a trajectory tracking control this does not necessarily hold, at least when the sensors and actuator are taken into account.

Ideal simulation: In order to further study the influence of the HGO on the trajectory tracking control, additionally an ideal simulation is performed for the eight trajectory, cp. Fig. 10.5 to 10.7.

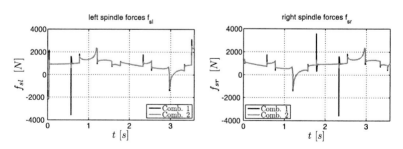

Figure 10.5: Spindle forces for Comb. 1 and 2 (ideal simulation)

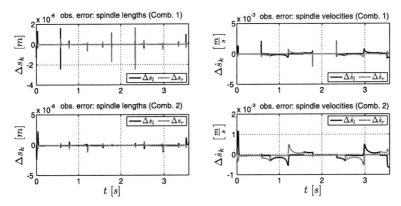

Figure 10.6: Observation errors in case of Comb. 1 and 2 (ideal simulation)

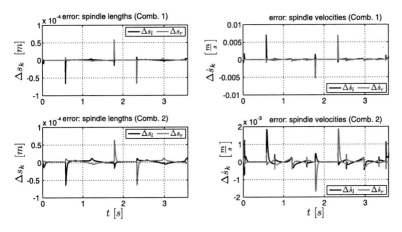

Figure 10.7: Tracking errors in case of Comb. 1 and 2 (ideal simulation)

For this case also Comb. 1 does not show a bang-bang behavior of the forces (cp. Fig. 10.5), as a consequence its performance is due to its higher gains better than that of Comb. 2.

Comparison and differentiating effect: Whereas in case of a realistic simulation the courses of the observed velocities (cp. Fig. 10.3) and the tracking errors

(cp. Fig. 10.4) are noticeably noisy – especially for the higher gains of Comb. 1 –
all errors are almost one decimal power smaller (cp. Fig. 10.6 and 10.7) in the ideal
simulation mode. The reason for the noisy plots above is the operating mode of the
sensors, which give a quantized information of the position changes. These jumps
in the signals lead to noise caused by the observers. The reason for it is the differentiating effect of a HGO, which means that for high gains its transfer-function
tends to that of a differentiator [DK97, Kha02]. Therewith a larger observation
error in case of a realistic simulation and in the results of the practical implementation can be explained: The stepwise change of s_k in each sampling step causes
peaks due to the differentiating effect of the HGO. The set of these peaks occurs
as noise in the plots of the trajectory tracking errors and the observed lengths
and velocities. Thereby a larger tracking error is caused. In the context of HGOs
this is called *peaking phenomenon* [Kha02]. Despite the noise, the trajectories $s_{k,d}$
are tracked also in case of the realistic simulation very exactly. Nevertheless the
performance slightly decreases.

Aspects of the different trajectories

In this paragraph the aspects of the different trajectories (in the sense of their
'underlying' paths) are discussed. Performing an ideal simulation, the transitions
between the straights and circular elements (eight) and that of the straights (tri-
angles) can be observed very well in the error plots in the form of jumps. The
trajectory have been planned (smooth) and time-optimal in Sec. 7.3. For that
reason the pseudo-velocity along the straights is considerably higher than on the
circular segments in case of the eight (cp. Fig. C.10). Physically the difference
can be explained by the fact, that along the straights the whole drive energy is
used for the driving, whereas in the other case a part of the energy is lost due to
the centrifugal force. For this reason, jumps in the profile of the pseudo-velocity
occur, which are interpreted as jumps of the reference value. These have to be
compensated by the tracking controller, leading to jumps in the error plots. The
jumps can also be observed at the transitions between the straights in case of the
triangle. In front of such a point the end-effector of the robot is braked to stand.
After the transition it is again accelerated. For that reason, at the transitions the
different smoothness levels (jerk limits) of the trajectories can be distinguished.
From Fig. 10.8 (ideal simulation) it is obvious that the jumps increase with the
jerk limit. The results given in Fig. 10.8 for the triangle, can be observed in the
same form for the eight. Furthermore, the influence of the jerk limit can also be
observed at the circular trajectories. No jumps occur, but an increasing jerk limit
leads to increasing tracking errors (cp. Tab. 10.9, ②, realistic simulation).
The results make also obvious that PCSMTOPTs lead to smaller tracking errors

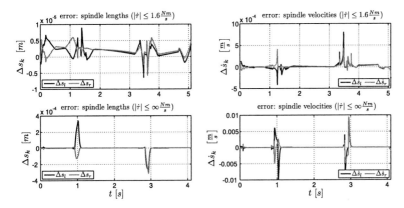

Figure 10.8: Triangle tracking errors for two different jerk levels (Comb. 2)

than pure time-optimal ones. Due to smooth trajectories the performance of the trajectory tracking controller is improved. Furthermore PCSMTOPTs prevent oscillations of the system. On the other hand the traveling time (slightly) increases.

Preview: The effect that the tracking errors are lower in case of trajectories planned for lower jerk limits can be observed performing an ideal or realistic simulation of Comb. 2. But in case of Comb. 1 (and almost all latter studied algorithms) the relationship is not valid performing a realistic simulation. In contrast at the demonstrator the observations are valid for all control algorithms in case of all trajectories, except for Comb. 2 in case of the circle trajectories. But since Comb. 1 and 2 are the same control structure it is correct to say, that for appropriate chosen controller (and observer) gains, the theoretical and experimental results concerning the positive effect of PCSMTOPTs on the tracking accuracy match.

10.3.3 Reference control with different velocity reconstruction

In the following the HGO is replaced by the three other algorithms derived in Chap. 9: the LKF, the end-fit-FOAW and best-fit-FOAW. As controller gains that of C2 are chosen. For a direct comparison with the HGO again a realistic simulation for the eight trajectory is performed.

Like in case of the HGO of Comb. 1 in all three cases the spindle-forces show bang-bang behavior (which is insignificantly less in case of the FOAW algorithms). Since no further information can be taken from their courses they are omitted in this section, whereas the plots of the observation and the tracking errors are given in Fig. 10.9[1] and Fig. 10.10, respectively. From them it becomes obvious that the errors in case of all three algorithms are of the same scale, whereas that of the end-fit algorithm are slightly worse, at least in case of the eight-trajectory. A further analysis is performed in Sec. 10.3.4 based on the experimental results.

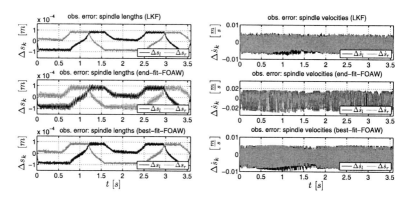

Figure 10.9: Observation errors in case of LKF, end-fit-FOAW and best-fit-FOAW

Moreover the windows lengths in case of the FOAW algorithms have to be studied. As obvious from Fig. 10.11 – like in case of the test scenario in Sec. 9.3.3 – they remain constant for both algorithms after they are established in the first steps. In this context the best-fit-FOAW has to be further studied. For a fixed filter length of $n = 2$, like here, it is identical to an end-fit-FOAW with the same fixed filter length. Since the end-fit algorithm has a lower computational burden, the performance of the tracking control can possibly be enlarged by applying it in the described way. But this approach stands in contrast to the basic idea of the AW algorithms and for different trajectories or changing conditions the considerations are maybe not valid anymore. The experimental results in Sec. 10.3.4 further confirm, that it is better to adapt n.

---

[1]In case of the FOAW algorithms there is not really an observation error in case of the position. Moreover its value is delayed until also the velocity value is calculated.

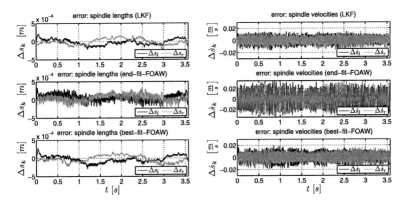

Figure 10.10: Tracking errors in case of LKF, end-fit-FOAW and best-fit-FOAW

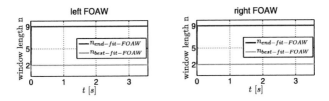

Figure 10.11: Windows lengths in case of the eight trajectory (gains C2)

10.3.4 Implementation at the demonstrator

In this section particularities observed at the demonstrator for tracking the trajectories using the two controller-observer combinations are summarized. For the implementation of the algorithms the dSPACE-system introduced in Sec. 2.2.4 is used. In contrast to the simulation, additionally a friction compensation, according to [BI05] has been implemented in case of the linearizing feedback.

Results for Comb. 1 and 2

From the measurement results at the demonstrator (cp. Tab. 10.10) it gets obvious that the performance of Comb. ① is in contrast to the simulations (cp. Tab. 10.10) better than that of Comb. ②. Also no rigorous bang-bang behavior of the actuator forces in case of Comb. 1 occurs. Whereas the drive degree of utilization is lower for

Comb. 1, it marginally increases in case of Comb. 2 compared to the simulations. The reason for the poorer performance of Comb. 2 seems to be founded in its low gains.

Overshoots

Like in case of the simulations (cp. Sec. 10.3.2) also at the demonstrator overshoots at transition points can be observed. This situation is exemplarily studied for the transitions of the straights in case of the triangle trajectories. Exemplarily the left spindle lengths and velocities in case of an unlimited jerk are given in Fig. 10.12 to Fig. 10.14 for the different controller-observer combinations. From Fig. 10.14 it becomes obvious, that at non-differentiable points of the velocity courses $\dot{s}_k$ overshoots occur.

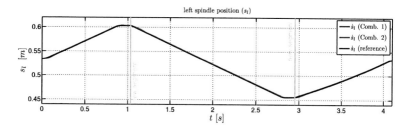

Figure 10.12: Left spindle length for triangle trajectory with $|\dot{\boldsymbol{\tau}}| \to \infty \frac{Nm}{s}$

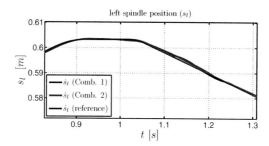

Figure 10.13: Zoom on Fig. 10.12

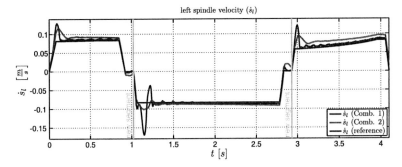

Figure 10.14: Left spindle velocities for triangle trajectory with $|\dot{\tau}| \to \infty \frac{Nm}{s}$

Also for the spindle length an increasing error in the region of these points can be observed (cp. Fig. 10.13). Furthermore (as already discussed above) the tracking errors of the spindle-positions and velocities are larger in case of Comb. 2.

Also the jerk limits of the trajectories have an influence on the overshoots. For analyzing this aspect the results for Comb. 1 for all three triangle trajectories are given in Fig. 10.15. As expected the overshoots (and tracking errors) increase with the jerk-limit.

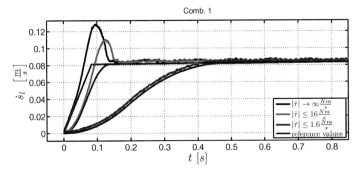

Figure 10.15: Left spindle velocities for the triangle trajectories (Comb. 1)

Run-up and end-point errors

In Fig. 10.16 exemplarily the left spindle positions of the triangle trajectories tracked with Comb. 1 are shown. As expected, the tracking control – not further optimized for the individual trajectories – works better in case of trajectories planned for a lower jerk limit. This does hold for the shown run-up and also for the mean values of all reference trajectories at the demonstrator.

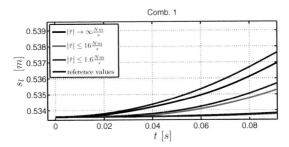

Figure 10.16: Zoom on the left spindle positions (beginning, Comb. 1)

At the end-points of the trajectories a further phenomenon can be observed. For analyzing this aspect the triangle trajectories are shifted in Fig. 10.17 to a common end-point (t_f). From the courses it becomes obvious that for higher jerk limits (and

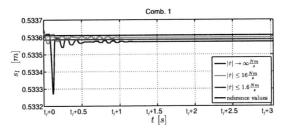

Figure 10.17: Zoom on the left spindle positions (ending, Comb. 1)

consequently more introduced energy by tracking the trajectories) the SpiderMill needs longer to get in its resting position.

Naturally the end-oscillation stands in a strong relationship with the chosen observer and controller gains. To have a further look on this point Fig. 10.18 is given.

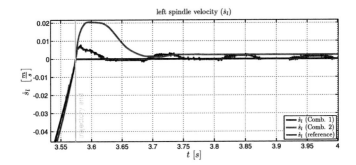

Figure 10.18: Left spindle velocities for the eight trajectory

There the left spindle velocities in case of the eight-trajectory are shown for the two combinations. Here, even better it can be seen, that the drives do not stand still at the end of the trajectory. In case of Comb. 1 it seems, that the spindle position oscillates between two increments and therefore the spindle velocity is 'swinging'. This is caused by the noise of the observer, increased by the coupling of the drives of the structure and amplified by the high controller gains. The many joints of the mechanism aid the oscillation. In case of Comb. 2 the influence is less due to the lower gains. But as a direct consequence also the derivation from the final value is lager.

10.3.5 Basic considerations for all control concepts

In this section the simulated and measured results for the reference tracking strategy including its extensions are discussed and some basic decisions for all further experiments are made.

High and low controller gains

For the realistic simulation (cp. ②, Tab. 10.9) it seems, that Comb. 2 shows a superior control performance based on lower controller and observer gains. But this does not necessarily hold in practice like the results in Sec. 10.3.4 indicate. The most important factor, limiting the controller performance is the bang-bang behavior of the control inputs. It has to be studied under several aspects: In case of the simulation environment the ball-screw-systems including the friction losses (non-linear component) of the drive systems are not modeled, whereas in case of the experimental realization the actuator dynamics is 'reduced/slowed' or even 'damped' (cp. Sec. 10.3.4) by them. This means, the bang-bang behavior will not occur at this degree at the physical demonstrator. Furthermore according to Sec. 10.3.3 also the observation concept in combination with the controller seems to have an impact on the actuator forces (cp. also Sec. 10.6).

At the demonstrator it can be expected, that to some degree higher gains (especially of the controller) will lead to a faster convergence of the errors and therefore to an enhanced performance. This assumption is confirmed by the results above and that in Tab. 10.10. But this point has also to be discussed with respect to the in practice very important aspects of energy and wear. If the tracking accuracy is high enough for a task these two points play the most important role and they are directly related to the drive degree of utilization. But in case of the SpiderMill the tracking errors are – at least in this thesis, only performing a position control – regarded as more important. Since they are in case of Comb. 1 significantly lower, its controller gains (C1) are used below in combination with the other observers.

Measurement of spindle length and friction compensation

In Chap. 9 studying the algorithms for the determination of the spindle velocities, also the spindle length are observed (at least in some cases) in order to verify the quality of the observation. But for the tracking control the measured length will be used directly. The only reasons for using the observed values instead, would be the cases of too much measurement noise or a too low resolution of the signals. But the results in Tab. 10.1 show the slight improvement by using the measured values.

The last main difference of the simulated and realized algorithms is the friction handling in case of the feedback linearization. For the simulation (where no friction is modeled) no compensation is necessary, whereas the control performance at the demonstrator is improved by compensating it – cp. Tab. 10.1.

Conditions	$\overline{\Delta P}$ $[m]$	$\overline{\Delta S}$ $[m/s]$	$\overline{\Delta \dot{S}}$ $[m/s^2]$	$\overline{\eta}$
Comb. 1, SO, NF	$1.5606 \cdot 10^{-4}$	$1.1965 \cdot 10^{-4}$	$2.1424 \cdot 10^{-3}$	49.0145
Comb. 1, SO, F	$1.4766 \cdot 10^{-4}$	$1.1666 \cdot 10^{-4}$	$1.9950 \cdot 10^{-3}$	49.1903
Comb. 1, SM, NF	$1.4972 \cdot 10^{-4}$	$1.1860 \cdot 10^{-4}$	$2.1274 \cdot 10^{-3}$	48.6814
Comb. 1, SM, F	$1.4381 \cdot 10^{-4}$	$1.1597 \cdot 10^{-4}$	$1.9984 \cdot 10^{-3}$	48.5721
Comb. 2, SO, NF	$1.7156 \cdot 10^{-3}$	$3.8685 \cdot 10^{-4}$	$6.3708 \cdot 10^{-3}$	35.0972
Comb. 2, SO, F	$1.5799 \cdot 10^{-3}$	$3.4768 \cdot 10^{-4}$	$5.7263 \cdot 10^{-3}$	35.1448
Comb. 2, SM, NF	$1.6293 \cdot 10^{-3}$	$3.7235 \cdot 10^{-4}$	$6.3226 \cdot 10^{-3}$	34.9788
Comb. 2, SM, F	$1.4916 \cdot 10^{-3}$	$3.3341 \cdot 10^{-4}$	$5.6448 \cdot 10^{-3}$	34.8624
SO/SM: observed/measured spindle length, (N)F: (no) friction compensation				

Table 10.1: Basic settings at the demonstrator

It might have been expected, that η would be slightly larger in case of friction compensation. But the results show, that this does not necessarily hold. The reason for it is the fact, that the lower tracking errors result in a lower control action.

As a consequence of these results in the following the measured spindle lengths will be used, the friction will be compensated and the controller gains C1 will be applied for studying the other observation algorithms in combination with the reference tracking control.

In Fig. 10.19 and 10.20 exemplarily the results of the practical simulation and the experimental ones are opposed for the eight trajectory. It becomes obvious that the scales of the tracking errors are identical, which indicates a very good modeling. The results for all other trajectories are summarized in Tab. 10.9 and 10.10.

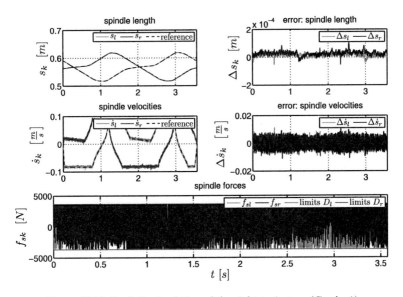

Figure 10.19: Realistic simulation of the eight trajectory (Comb. 1)

Tuning of the LKF

In Sec. 9.2.2 the performance of the LKF in dependency of Q has been studied in case of a realistic simulation (gains C2) also at the demonstrator (gains C1) its influence has to be further analyzed. The measurement results summarized in

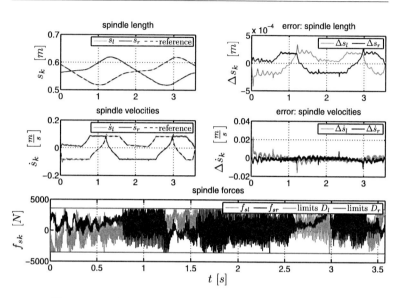

Figure 10.20: Measurements for the eight trajectory (Comb. 1)

Tab. 10.2 justify to chose $Q = 6 \cdot 10^{-2}$ also in case of the experimental setup in the following.

Q	$\overline{\Delta P}$ [m]	$\overline{\Delta S}$ [m/s]	$\overline{\Delta \dot{S}}$ [m/s²]	$\overline{\eta}$
$7 \cdot 10^{-3}$	$9.4111 \cdot 10^{-5}$	$1.0778 \cdot 10^{-4}$	$1.7172 \cdot 10^{-3}$	67.1647
$6 \cdot 10^{-2}$	$8.0714 \cdot 10^{-5}$	$1.0504 \cdot 10^{-4}$	$1.7701 \cdot 10^{-3}$	68.4568
1	$8.0734 \cdot 10^{-5}$	$1.0497 \cdot 10^{-4}$	$1.7901 \cdot 10^{-3}$	68.5512

Table 10.2: Experimental results for different Qs (gains C1, friction compensation)

Velocity reconstruction with FOAW

Also in case of the experimental results the adaption process of the windows length n is studied. Like the results in Fig. 10.21 (exemplarily given for the eight trajectory) show, in contrast to the simulations, n is changing for the best-fit-FOAW. The length in case of the end-fit-FOAW remains constant. In case of the Spider-Mill no limiting computational burden for the best-fit algorithm can be observed.

In case of simple control strategies – like the non-model-based below – normally

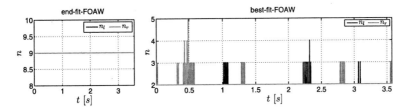

Figure 10.21: Windows length n for the eight trajectory (C1)

also basic concepts for the derivation of required velocities are realized. Especially, when the hardware resources are limited algorithms like the above introduced end- and best-fit-FOAW are applied or even numerical differentiations are made. In order to accommodate this aspect in this thesis, based on the results of Tab. 10.10, in all further cases, where no specific velocity observation is part of the control concept itself (cp. LQG, Sec. 10.4.1) the best-fit-FOAW is used due to its better performance. This does also hold for the corresponding simulations.

One further point has to be mentioned. The bad results in case of the simulation of the end-fit-FOAW would not occur in case of the controller gains C2 – cp. Sec. 10.3.3. An implementation error can also due to the experimental results be excluded. Since they show an acceptable performance also in case of the end-fit-FOAW no further investigations are made.

10.4 Further model-based control strategies

In this section further model-based control concepts commonly known in robotic literature are presented. Except the last one, all of them are directly related to the reference strategy, basing on the feedback-linearized system. In contrast to the reference control whose parameter choice is more or less heuristic these strategies for the linearized system feature a more systematic design.

10.4.1 LQG control

The LQG (*Linear Quadratic Gaussian*) control is the combination of a LKF (a *Linear Quadratic Estimator* - LQE) with a *Linear Quadratic Regulator* (LQR). Traditionally, it is assumed that the plant dynamics is linear and known. Furthermore also the statistical properties of the noise processes are assumed as known [SP05]. This implies that the model (9.19) is valid. Due to the separation principle the

LKF and LQR can be designed and computed independently.

Theory of LQR

The theory for the LQR and its calculation steps are well known in literature [FPW90, Ż03, SP05] and are for that reason only shortly sketched below.
Starting from the discrete system model (9.19) the goal of the LQR design is to find the optimal control law

$$\mathbf{u}_k = -\mathbf{K}_{r,k}^T \cdot \mathbf{x}_k, \tag{10.16}$$

which minimizes a quadratic cost-function

$$J = \frac{1}{2} \sum_{k=0}^{N} \left[\mathbf{x}_k^T \mathbf{Q}_r \mathbf{x}_k + \mathbf{u}_k^T \mathbf{R}_r \mathbf{u}_k \right], \tag{10.17}$$

where $\mathbf{Q}_r = \mathbf{Q}_r^T \geq 0$ and $\mathbf{R}_r = \mathbf{R}_r^T > 0$ are constant and appropriately chosen weighting matrices. In most cases they are chosen as diagonal matrices with all elements positive or zero (only for $\mathbf{Q}_r$). In order to solve the problem defined above it is normally reformulated into a standard constrained minimization problem, which is solved by using the method of Lagrange multipliers. The solution steps can e.g. be found in the [FPW90]. Finally, the desired optimal time-varying feedback gain matrix $\mathbf{K}_{r,k}$ is obtained. No knowledge of the initial system state is required and the optimal gains $(\mathbf{K}_{r,0}, \dots \mathbf{K}_{r,N})$ can be pre-computed.
But since in case of the algorithm with finite horizon the gain matrix $\mathbf{K}_{r,N}$ in the last point of the trajectory is zero, the final position is not kept. For realizing a safe operation mode a switching to another algorithm would be necessary. Therefore, the steady-state (infinite horizon, $N \to \infty$) solution $\mathbf{K}_r$ of the LQR-problem is applied in the following [Nai03]. Since in the *Matlab/Control System Toolbox* an algorithm for its calculation is directly available, the *LQRD* command is used for the design.

The LQG control concept

Combining the optimal sate estimation and the optimal state feedback the structure shown in Fig. 10.22 results. In order to distinguish the gain matrices $\mathbf{K}_f$ is used for the LKF and $\mathbf{K}_r$ for the LQR. The discrete-time representation of the LQG is given by

$$\mathbf{x}_k = \left(\mathbf{A}_{k-1} - \mathbf{B}_{k-1} \mathbf{K}_r^T \right) \mathbf{x}_{k-1} + \mathbf{B}_{k-1} \mathbf{K}_r^T \mathbf{e}_{x,k-1} + \mathbf{w}_{k-1}, \tag{10.18}$$

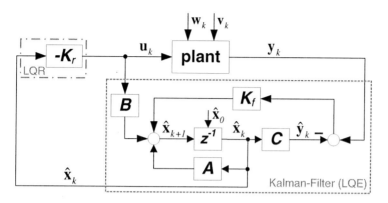

Figure 10.22: LQG control structure

where the state estimation error $\mathbf{e}_{x,k}$ is defined to

$$\mathbf{e}_{x,k} = \mathbf{x}_k - \hat{\mathbf{x}}_k = \left(\mathbf{A}_{k-1} - \mathbf{K}_{f,k-1}^T \mathbf{C}_{k-1}\right) \mathbf{e}_{x,k-1} + \mathbf{w}_{k-1} - \mathbf{K}_{f,k-1}^T \mathbf{v}_{k-1}. \quad (10.19)$$

Consequently, for the plant and the LQG controller, according to Fig. 10.22 the closed loop dynamics is given by [SP05]

$$\begin{bmatrix} \mathbf{x}_k \\ \mathbf{x}_k - \hat{\mathbf{x}}_k \end{bmatrix} = \begin{bmatrix} \mathbf{A}_{k-1} - \mathbf{B}_{k-1}\mathbf{K}_r^T & \mathbf{B}_{k-1}\mathbf{K}_r^T \\ \mathbf{0} & \mathbf{A}_{k-1} - \mathbf{K}_{f,k-1}^T \mathbf{C}_{k-1} \end{bmatrix} \begin{bmatrix} \mathbf{x}_{k-1} \\ \mathbf{x}_{k-1} - \hat{\mathbf{x}}_{k-1} \end{bmatrix}$$
$$+ \begin{bmatrix} \mathbf{I} & \mathbf{0} \\ \mathbf{I} & -\mathbf{K}_{f,k-1}^T \end{bmatrix} \begin{bmatrix} \mathbf{w}_{k-1} \\ \mathbf{v}_{k-1} \end{bmatrix}. \quad (10.20)$$

Eq. (10.20) confirms that the separation principle is valid for the LQG, which allows a separate design of the LKF and the LQR. The steady state gain-matrix $\mathbf{K}_r$ is calculated offline.

Implementation of the LQG at the SpiderMill

The control structure of Fig. 10.22 is applied to the feedback-linearized SpiderMill. The advantage of the LQR is that the controller gain matrix is determined in an optimal way, based on a cost criterion. In contrast the gains of the reference control have been chosen heuristically. The limitations of the actuator forces are not considered explicitly in form of an inequality, but can be considered indirectly by an appropriate choice of the weighting matrix $\mathbf{R}_r$.

The matrices $\mathbf{Q}_r$ and $\mathbf{R}_r$ are chosen in a way to simultaneously consider the error dynamics and the limitation of the actuator forces. On the one hand the error dynamics should converge as fast as possible and on the other hand, the actuator forces should not saturate or show a bang-bang characteristic (if not only a high tracking accuracy is the goal – like below).

The for three different weighting matrices (cp. Tab. 10.3) calculated gains (cp. Tab. 10.3) have been implemented at the demonstrator, yielding the performance indices summarized in Tab. 10.4.

variant	$\mathbf{Q}_r$	$\mathbf{R}_r$	$\mathbf{K}_r$
variant 1	$0.5\,\mathrm{diag}((1/0.011)^2, (1/0.11)^2)$	$(1/20)^2$	$[1201.4110,\ 129.7563]^T$
variant 2	$0.5\,\mathrm{diag}((1/0.00015)^2, (1/0.13)^2)$	$(1/20)^2$	$[75471.7297,\ 398.1543]^T$
variant 3	$0.5\,\mathrm{diag}((1/0.00014)^2, (1/0.13)^2)$	$(1/20)^2$	$[80269.5814,\ 409.8924]^T$

Table 10.3: Weighting and gain matrices for the LQR design

variant	$\overline{\Delta P}\ [m]$	$\overline{\Delta S}\ [m/s]$	$\overline{\Delta \dot{S}}\ [m/s^2]$	$\overline{\eta}$
variant 1	$3.0832 \cdot 10^{-3}$	$5.9544 \cdot 10^{-4}$	$1.8221 \cdot 10^{-3}$	34.4841
variant 2	$8.0552 \cdot 10^{-5}$	$1.0584 \cdot 10^{-4}$	$1.6449 \cdot 10^{-3}$	57.0141
variant 3	$7.7241 \cdot 10^{-5}$	$1.0504 \cdot 10^{-4}$	$1.7929 \cdot 10^{-3}$	58.7764

Table 10.4: Performance indices for the LQGs

From the choice of the weighting matrices the here performed prioritization of the tracking errors becomes obvious: Both the position and the velocity tracking errors are weighted stronger than the control inputs (actuator forces). Furthermore the position tracking error is regarded as more important than the velocity tracking error (whereas both are interconnected).

Comparing the experimental results of variant 3 (ⓒ) with that obtained applying the LKF together with the gains C1 (③ in Tab. 10.10) it can be observed that even lower but more systematically chosen gains lead to an enhanced performance.

10.4.2 Loop shaping approach

In this section a further linear controller matrix $\mathbf{G}_C(s) = [G_{C,1}(s),\ G_{C,2}(s)]^T$ is designed for the feedback-linearized system $\mathbf{G}(s) = [G_1(s),\ G_2(s)]^T$ applying the Loop Shaping approach. The control structure is given in Fig. 10.23. The spindle velocities required for the linearization are determined using the Best-Fit-FOAW.

The theoretical background of the approach, recapitulated below, is mainly taken from [SP05]. In case of SISO systems loop shaping means the forming of the amplitude characteristics and the phase response. The specifications are defined for the frequency response of the open loop $L(s) = G_C(s) \cdot G(s)$. In case of MIMO systems the demands are for the courses of the singular values. The specifications are done in a way that the closed loop has a good performance and robustness. For simple plants a beneficial course of $L(s)$ can be achieved with simple linear controllers, whereas in case of complex systems, which e.g. crosses the 0 dB line several times an H_∞ approach is necessary [MG92]. Since in case of the SpiderMill it can be assumed that the feedback-linearized system has decoupled I^2 behavior, two identical controllers $G_{C,i}(s)$ $(i = 1, 2)$ are designed based on the Bode plot.

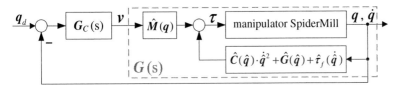

Figure 10.23: Control structure for the Loop Shaping approach

Specifications for the open loop

In a first step the relationship between the open $L(s)$ and closed loop $T(s)$ is shortly summarized. Commonly the number of poles of $L(s)$ is larger than the number of zeros, which implies $\lim_{s\to\infty} L(s) = 0$. Considering the relationship

$$T(s) = \frac{L(s)}{1 + L(s)} \quad \text{with} \quad L(s) = G_C(s) \cdot G(s) \qquad (10.21)$$

the following approximation is valid for high frequencies

$$T(s) \approx L(s), \ \omega \gg 0, \qquad (10.22)$$

For small frequencies $T(s) \approx 1$ should hold. This can only be achieved if $L(s)$ is very large in this frequency range. Based on these considerations two fundamental requirements on the course of $L(s)$ can be defined. For a good tracking behavior, $L(s)$ has to be very large in the lower frequency range. In contrast, for the suppression of disturbances $L(s)$ should be small for high frequencies and decrease with 40 dB per decade. If the open loop shows this characteristic, the closed loop has PT_2 behavior. Further important parameters are the crossover frequency ω_c, which determines the bandwidth of the closed loop (a larger bandwidth leads to a faster control) and the slope of the amplitude characteristics of $L(s)$. The slope of $L(s)$ for a given frequency ω_x is for minimum-phase systems directly related to the phase response. The following relationship holds [SP05]:

$$\angle L(j\omega_x) = \frac{1}{\pi} \int_{-\infty}^{\infty} \underbrace{\frac{d\ln|L(j\omega)|}{d\ln\omega}}_{N(\omega)} \cdot \ln\left|\frac{\omega+\omega_x}{\omega-\omega_x}\right| \cdot \frac{d\omega}{\omega}, \qquad (10.23)$$

which can in case of stable minimum-phase systems be approximated by [SP05]

$$\angle L(j\omega_x) \approx \frac{\pi}{2} \cdot N(\omega_x)\,[\mathrm{rad}] \mathrel{\widehat{=}} 90° \cdot N(\omega_x) \qquad (10.24)$$

The approximation is exact for systems of the form $G(s) = 1/s^n$ (where the term $N(\omega) = -n \to \angle L(j\omega) = -n \cdot 90°$) and good for stable minimum-phase systems except at frequencies close to those of complex poles or zeros [SP05]. Since in case of the SpiderMill the subsystems are double integrators (ideal case), it is assumed that it is exact.

Consequently in the area of the crossover frequency ω_c the slope of $L(s)$ is significant for the phase margin φ_r. A slope of -40 dB per decade, which corresponds to a double integrator, would lead to a marginally stable system. This is undesired. But for small frequencies ($\omega \ll \omega_c$) a slope of -40 dB per decade is beneficial, since it allows a good tracking of ramps as reference inputs. Also in the high frequency range ($\omega \gg \omega_c$) a slope of -40 dB per decade is necessary in order to yield a good suppression of disturbances.

Application to the feedback-linearized model of the SpiderMill

Since the subsystems $G_i(s)$ ($i = 1, 2$) obtained by the linearization have I^2 behavior over the whole frequency range, $G_i(s)$ decreases with -40 dB per decade and the phase shift is $-180°$. The gain response crosses the 0 dB line at $\omega_c = 1$ rad/s. The goal of the controller design is to fulfill the above defined specification. In order to get a large bandwidth, a high steady state amplification is necessary. Furthermore the slope of $L(s)$ has to be reduced to some value around -20 dB in the region of ω_c. In order to fulfill the specifications three lead compensators are tested (identical zero and pole locations; different gains). The corresponding experimental results are summarized in Tab. 10.5.

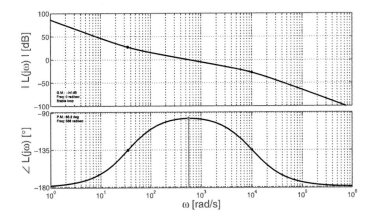

Figure 10.24: Bode-diagram of the open loop $L(s)$ in case of variant 3

variant	φ_r [°]	ω_c [rad/s]	$\overline{\Delta P}$ [m]	$\overline{\Delta S}$ [m/s]	$\overline{\Delta \dot{S}}$ [m/s²]	$\overline{\eta}$
variant 1	79.1	$2.03 \cdot 10^2$	$5.929 \cdot 10^{-4}$	$1.912 \cdot 10^{-4}$	$1.514 \cdot 10^{-3}$	40.35
variant 2	81.5	$1.2 \cdot 10^3$	$2.417 \cdot 10^{-4}$	$1.283 \cdot 10^{-4}$	$3.262 \cdot 10^{-3}$	95.84
variant 3	83.2	$5.68 \cdot 10^2$	$2.007 \cdot 10^{-4}$	$1.253 \cdot 10^{-5}$	$1.264 \cdot 10^{-3}$	76.69

Table 10.5: Design of the loop shaping controller

The best results (variant 3) have been obtained for the following, discretely realized compensator $\mathbf{G}_C(z) = [G_{C,1}(z),\ G_{C,2}(z)]^T$ with

$$G_{C,i}(z) = \frac{1.1546 \cdot 10^{10} z - 1.1149 \cdot 10^{10}}{12000 z + 8000} \quad i = 1, 2. \tag{10.25}$$

From the Bode plot in Fig. 10.24 it becomes obvious that the specifications have been reached. The large phase margin implies a low overshoot and a robust stability. Although a higher bandwidth leads to a faster closed-loop control, the results in Tab. 10.5 indicate that in case of the here performed tracking control a larger phase margin is even more important.

10.4.3 PD control with gravity compensation

A controller, originally developed for the tracking of constant set-points, is the PD controller with gravity compensation [TA81]. Several different versions have been introduced in the past, which can be distinguished mainly by the point whether reference (d) values are used for calculating $\mathbf{G}(\mathbf{q}_{(d)})$ or not. The strategy applied here is given in Fig. 10.25. Based on Lyapunov theory it can be proven that for any positive definite controller gain matrices $\mathbf{K}_P$ and $\mathbf{K}_D$ the closed loop is globally asymptotically stable. The controller can also be used in case of tracking-problems, where also the stability is ensured. A detailed stability analysis can be found in [Kel97]. There additionally the robustness in case of parameter uncertainties is discussed and an adaptive version of the controller is presented.

For the control structure in Fig. 10.25 the following control law is valid

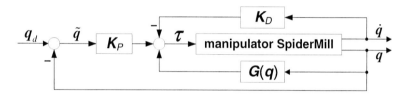

Figure 10.25: PD control with gravity compensation

$$\boldsymbol{\tau} = \mathbf{K}_P\tilde{\mathbf{q}} - \mathbf{K}_D\dot{\mathbf{q}} + \mathbf{G}(\mathbf{q}), \qquad (10.26)$$

where $\mathbf{K}_P$ and $\mathbf{K}_D$ are $n \times n$ symmetric positive-definite matrices and the joint position error is defined by $\tilde{\mathbf{q}} = \mathbf{q}_d - \mathbf{q}$. The index d stands for the desired joint coordinates. Another class of existing PD controllers with gravity compensation apply the following control structure

$$\boldsymbol{\tau} = \mathbf{K}_P\tilde{\mathbf{q}} - \mathbf{K}_D\dot{\tilde{\mathbf{q}}} + \mathbf{G}(\mathbf{q}) \quad \text{with} \quad \dot{\tilde{\mathbf{q}}} = \dot{\mathbf{q}}_d - \dot{\mathbf{q}}. \qquad (10.27)$$

Strategies for choosing the controller gain matrices can be found e.g. in [Tom91, Kel97,EHE08]. For the choice of $\mathbf{K}_P$ the following relationship is applied here [Kel97]

$$\lambda_m\left\{\mathbf{K}_P\right\} \geq n\left(\max_{i,j,\mathbf{q}}\left|\frac{\partial g_i(\mathbf{q})}{\partial q_j}\right|\right) \quad \text{with} \quad i,j = 1\ldots 2, \qquad (10.28)$$

where $\lambda_m\left\{\mathbf{K}_P\right\}$ is the smallest eigenvalue of the symmetric positive-definite bounded matrix $\mathbf{K}_P$, $g_i(\mathbf{q})$ the ith element of the vector $\mathbf{G}$ and n the dimension of $\mathbf{q}$.

When computed torque control concepts are used in combination with a PD controller of the structure (10.27), it is possible to set up the following linear relationship for the error

$$\ddot{\mathbf{e}} + \mathbf{K}_D \dot{\mathbf{e}} + \mathbf{K}_P \mathbf{e} = 0. \qquad (10.29)$$

Since in case of the SpiderMill the controllers for both sides of the robot are chosen identical, in the following a scalar notation is used. The solution of Eq. (10.29) yields a second order damped system with the natural frequency ω_0 and the damping ratio d

$$\omega_0 = \sqrt{k_P} \qquad \text{and} \qquad d = \frac{k_D}{2\sqrt{k_P}}, \qquad (10.30)$$

where ω_0 determines the speed of the system response. In robotics it is customary to take $d = 1$ leading to a critically damped response, which is the fastest non-oscillatory one [EHE08]. So the value of k_D should be chosen to

$$k_D = 2\sqrt{k_P}. \qquad (10.31)$$

Even though the control structure (10.26) is implemented, the rules (10.28) and (10.31) give a good hint for the first choice of k_P and k_D. But a (fine-)tunning is still necessary in order to enhance the control performance. After a further tunning of the factors at the demonstrator k_P is chosen to $k_P = 1\,200\,000$ and k_D according to Eq. (10.31).

10.5 Non-model-based control concepts

In the following a decentralized and a centralized non-model-based control strategy are introduced. In order to be consistent to the corresponding models of Chap. 5 used for their derivation, also in this section the concepts are expressed in terms of the pseudo-active joint coordinates. But in case of the simulation and the experimental realization of the concepts the active joint coordinates are used instead. Required velocities are determined based on the best-fit-FOAW.

10.5.1 Single axis PID control (decentralized control)

In this section the single link with end-mass model of the SpiderMill is used for the left and right side of the robot in oder to design two identical PID controllers.

For this purpose Eq. (5.16) is solved for $\ddot{\theta}_k$

$$\ddot{\theta}_k = \frac{\frac{f_k\, d_{Ak}\, d_{sk}\, \sin(\theta_k + \theta_{offk})}{\sqrt{d_{Ak}^2 + d_s^2 - 2d_A d_s \cos(\theta_k + \theta_{offk})}} - g\,(d_{m1}\, m_1 + d_1\, m^*)\cos\theta_k}{I_{m12} + d_{m1}^2\, m_1 + d_1^2\, m^*} \tag{10.32}$$

Performing a linearization around a working point WP the following state-space model is obtained

states: $\Delta x_1 = \Delta\theta_k$ input: $\Delta u = \Delta f_k$ output: $\Delta y = \Delta x_1$
$\qquad\quad \Delta x_2 = \Delta\dot{x}_1$

model: $\begin{bmatrix} \Delta\dot{x}_1 \\ \Delta\dot{x}_2 \end{bmatrix} = \underbrace{\begin{bmatrix} 0 & 1 \\ k_1(\theta_k) & 0 \end{bmatrix}}_{\mathbf{A}} \begin{bmatrix} \Delta x_1 \\ \Delta x_2 \end{bmatrix} + \begin{bmatrix} 0 \\ k_2(\theta_{k,WP}) \end{bmatrix}\Delta u + \begin{bmatrix} 0 \\ k_3(\theta_{k,WP}, f_{k,WP}) \end{bmatrix}$

$\qquad\quad \Delta y = \begin{bmatrix} 1 & 0 \end{bmatrix}\begin{bmatrix} \Delta x_1 \\ \Delta x_2 \end{bmatrix}.$

$$\tag{10.33}$$

In order to develop a controller for the worst case, the largest pair of eigenvalues of $\mathbf{A}$ for all trajectories is determined to $\lambda_{1/2} = \pm 4.1746$. The corresponding WP is $x_1 = 1.7517$ rad and $x_2 = -5.1376 \cdot 10^{-6}$ rad/s, yielding the transfer function

$$G_{SA}(s) = \frac{8.882 \cdot 10^{-16}s + 0.003749}{s^2 - 8.882 \cdot 10^{-16}s - 17.43}. \tag{10.34}$$

This model is extended by a drive model, for which in order to derive only one controller the mean values of the parameters of Eq. (2.2) are applied.

Furthermore in order to allow a better interpretation and implementation of the control strategy the term $k_3(\theta_{k,WP}, f_{k,WP})$ is solved for the force $f_{k,WP} = k_4(\theta_{k,d}, \ddot{\theta}_{k,d})$ using reference values for $\theta_{k,d}$ and $\ddot{\theta}_{k,d}$. The implemented control structure is given in Fig. 10.26. The designed controller is given by

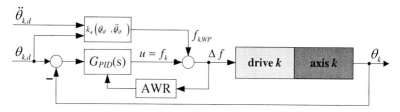

Figure 10.26: Single axis control

$$G_{PID}(s) = 2.3483 \cdot 10^7 \, \frac{(s + 0.991)\,(s + 5.179)}{s\,(s + 418.4)}. \qquad (10.35)$$

For this controller a phase margin of $\varphi_r = 75.7°$ at $\omega_c = 781$ rad/s is obtained. The step response of the closed-loop has a rise-time of $t_r = 28.8$ ms and a settling time of $t_{set} = 1.21$ s. By implementing the control law in sum representation also an AWR (Anti-Windup-Reset) for the integral part has been introduced.

10.5.2 State-space control (centralized control)

Derivation of the control structure

For developing a state-space control strategy for the SpiderMill, its equations of motion are linearized around several working points (WP). Starting form Eq. (5.23)

$$\mathbf{M}(\boldsymbol{\theta}) \cdot \ddot{\boldsymbol{\theta}} + \mathbf{C}(\boldsymbol{\theta}) \cdot \begin{bmatrix} \dot{\theta}_l^2 \\ \dot{\theta}_r^2 \\ \dot{\theta}_l \cdot \dot{\theta}_r \end{bmatrix} + \mathbf{G}(\boldsymbol{\theta}) = \boldsymbol{\tau} = \mathbf{f}(\boldsymbol{\theta}, \dot{\boldsymbol{\theta}}, \ddot{\boldsymbol{\theta}}), \qquad (10.36)$$

the strategy in [KES04] is applied to approximate the torque $\boldsymbol{\tau}$ by a Taylor series

$$\boldsymbol{\tau} \cong \mathbf{f}_{WP}(\boldsymbol{\theta}_{WP}, \dot{\boldsymbol{\theta}}_{WP}, \ddot{\boldsymbol{\theta}}_{WP}) + \mathbf{J}_1 \cdot \Delta\ddot{\boldsymbol{\theta}} + \mathbf{J}_2 \cdot \Delta\dot{\boldsymbol{\theta}} + \mathbf{J}_3 \cdot \Delta\boldsymbol{\theta}, \qquad (10.37)$$

where the $\mathbf{J}_i$s are Jacobian matrices. For the state-space representation $\mathbf{x}_1 = \boldsymbol{\theta} = \begin{bmatrix} \theta_l, & \theta_r \end{bmatrix}^T$ and $\mathbf{x}_2 = \dot{\boldsymbol{\theta}}$ are chosen as states. The WPs are defined along the end-effector paths. Since the joint velocities play no role for the determination of the WPs $\mathbf{x}_{2,WP} = 0$ holds, yielding $\Delta\boldsymbol{\theta} = \Delta\mathbf{x}_1 = \mathbf{x}_1 - \mathbf{x}_{1,WP}$ and $\Delta\dot{\boldsymbol{\theta}} = \Delta\mathbf{x}_2 = \mathbf{x}_2$. Furthermore also $\dot{\mathbf{x}}_1 = \mathbf{x}_2$ and $\Delta\dot{\mathbf{x}}_1 = \Delta\mathbf{x}_2$ hold. Since all WPs are time-independent, $\Delta\dot{\mathbf{x}}_1 = \dot{\mathbf{x}}_1 - \dot{\mathbf{x}}_{1,WP} = \dot{\mathbf{x}}_1$ and $\Delta\dot{\mathbf{x}}_2 = \dot{\mathbf{x}}_2$ are valid. This allows to express the linearized model (10.37) using absolute coordinates

$$\begin{bmatrix} \dot{\mathbf{x}}_1 \\ \dot{\mathbf{x}}_2 \end{bmatrix} = \underbrace{\begin{bmatrix} \mathbf{0} & \mathbf{I} \\ -\mathbf{J}_1^{-1}\mathbf{J}_3 & -\mathbf{J}_1^{-1}\mathbf{J}_2 \end{bmatrix}}_{\mathbf{A}} \underbrace{\begin{bmatrix} \mathbf{x}_1 \\ \mathbf{x}_2 \end{bmatrix}}_{\mathbf{x}} + \underbrace{\begin{bmatrix} \mathbf{0} \\ \mathbf{J}_1^{-1} \end{bmatrix}}_{\mathbf{B}} \boldsymbol{\tau} + \underbrace{\begin{bmatrix} \mathbf{0} \\ -\mathbf{J}_1^{-1}\mathbf{f}_{WP} \end{bmatrix}}_{\boldsymbol{g}} + \underbrace{\begin{bmatrix} \mathbf{0} \\ \mathbf{J}_1^{-1}\mathbf{J}_3\mathbf{x}_{1,WP} \end{bmatrix}}_{\boldsymbol{w}}$$

$$(10.38a)$$

$$\mathbf{y} = \underbrace{\begin{bmatrix} \mathbf{I} & \mathbf{0} \end{bmatrix}}_{\mathbf{C}} \begin{bmatrix} \mathbf{x}_1 \\ \mathbf{x}_2 \end{bmatrix}. \qquad (10.38b)$$

This representation is useful since trajectory tracking – and not fixed set-point – control is the goal. Since $\dot{\boldsymbol{\theta}}_{WP} = 0$ and $\ddot{\boldsymbol{\theta}}_{WP} = 0$ for all WPs, from Eq. (10.36) and (10.37) it follows that $\mathbf{f}_{WP} = \mathbf{G}(\boldsymbol{\theta}_{WP})$. Due to that reason the gravitational force only influences the dynamics by $\mathcal{G}$. With the feedforward compensation

$$\mathbf{F} = -(\mathbf{B}^T \cdot \mathbf{B})^{-1} \cdot \mathbf{B}^T \cdot \mathcal{G} \qquad (10.39)$$

its influence is compensated in the WPs[2]. Eq. (10.39) depends only on the WPs but not on the system states. Applying (10.39) the system has in each WP a rest position. Since the rest position is in contrast to a linear system in a WP not equal zero, the system is called affine. An affine system is a linear system with a coordinate displacement form the origin to the WP. This displacement is expressed by $\mathcal{W}$. The system structure is marked in Fig. 10.27.

From an eigenvalue analysis of the substitution system with feedforward compensation for the planned path it becomes obvious that the eigenvalues are only slightly changing. For that reason only a few WPs are required. Since for the state-space control the knowledge of the full system state is required, the velocities are determined based on the best-fit-FOAW.

In order to attain steady state accuracy the state-space control is extended by an integral state feedback. For that purpose the system output $\mathbf{y}$ is compared to its reference value $\mathbf{y}_d$ and the control error is added up. The integrator output $\mathbf{x}_{int}$ extends the state vector. The final structure is given in Fig. 10.27 additionally realizing an AWR. The controller is defined by $\mathbf{K} = \begin{bmatrix} \mathbf{K}_S & \mathbf{K}_{int} \end{bmatrix}$, where $\mathbf{K}_S$ is the gain of the control error of the states and $\mathbf{K}_{int}$ that of the integral error. The control law is given by

$$\boldsymbol{\tau} = \mathbf{K} \cdot \begin{bmatrix} \mathbf{x}_d - \mathbf{x} \\ \mathbf{x}_{int} \end{bmatrix} + \mathbf{F}. \qquad (10.40)$$

Applying the control law to the system (10.38) yields

$$\begin{bmatrix} \dot{\mathbf{x}} \\ \dot{\mathbf{x}}_{int} \end{bmatrix} = \begin{bmatrix} \mathbf{A} - \mathbf{B} \cdot \mathbf{K}_S & \mathbf{B} \cdot \mathbf{K}_{int} \\ \mathbf{C} & 0 \end{bmatrix} \cdot \underbrace{\begin{bmatrix} \mathbf{x} \\ \mathbf{x}_{int} \end{bmatrix}}_{\mathbf{x}_e} + \begin{bmatrix} \mathbf{B} \cdot \mathbf{K}_S \\ -\mathbf{C} \end{bmatrix} \cdot \mathbf{x}_d + \begin{bmatrix} \mathcal{W} \\ 0 \end{bmatrix} \qquad (10.41a)$$

$$\mathbf{y} = \begin{bmatrix} \mathbf{C} & 0 \end{bmatrix} \cdot \begin{bmatrix} \mathbf{x} \\ \mathbf{x}_{int} \end{bmatrix} \qquad (10.41b)$$

for the closed loop.

Implementation of the concept at the SpiderMill

The values of the controller gain matrices $\mathbf{K}(j)$ for the different WPs are determined based on an LQR design using the Matlab *LQRD* command. Additionally

---

[2]The feedforward compensation can be interpreted as an additive part of the control signal – cp. Eq. (10.40).

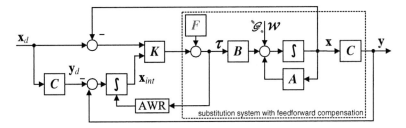

Figure 10.27: Block diagram of the state-space control

the individual feedforward compensations are calculated for the WPs. As switching condition the shortest distance between the TCP position and the WPs is used

$$\boldsymbol{\tau} = \mathbf{K}(j) \cdot \mathbf{x}_c + \mathbf{F}(j) \tag{10.42a}$$

$$j = \left\{ i \,\middle|\, \min_{1...n} \left((x(\boldsymbol{\theta}) - x_{WP}(i))^2 + (y(\boldsymbol{\theta}) - y_{WP}(i))^2 \right) \right\}, \tag{10.42b}$$

where n is the number of the WPs.

In order to evaluate the performance of the strategy three combinations of weighting matrices, summarized in Tab. 10.6, have been studied, yielding the performance indices summarized in Tab. 10.7 at the demonstrator.

variant	Q	R
variant 1	diag$(1900000, 1900000, 1350, 1350, 10, 10)$	diag$(0.001, 0.001)$
variant 2	diag$(28000000, 28000000, 1350, 1350, 10, 10)$	diag$(0.001, 0.001)$
variant 3	diag$(99000000, 99000000, 1350, 1350, 10, 10)$	diag$(0.001, 0.001)$

Table 10.6: Weighting matrices of the state-space controllers

variant	$\overline{\Delta P}$ [m]	$\overline{\Delta S}$ [m/s]	$\overline{\Delta \dot{S}}$ [m/s²]	$\overline{\eta}$
variant 1	$6.1938 \cdot 10^{-3}$	$1.2344 \cdot 10^{-3}$	$2.8204 \cdot 10^{-3}$	40.4107
variant 2	$4.2433 \cdot 10^{-3}$	$8.7565 \cdot 10^{-4}$	$2.3421 \cdot 10^{-3}$	40.0803
variant 3	$2.8195 \cdot 10^{-3}$	$6.1234 \cdot 10^{-4}$	$2.0083 \cdot 10^{-3}$	40.1781

Table 10.7: Performance indices of the state-space control

Comparing these results with that obtained for the single axis PID control (⊙ in Tab. 10.10) two points become obvious: In case of the trajectory tacking control of the SpiderMill higher controller gains of the state-space controller lead to a better performance (the gains are increasing from variant 1 to 3) and a (good tuned) centralized control strategy is always better than a standard decentralized one in case of parallel robots.

10.6 General results for the control strategies

In the following two tables (Tab. 10.9 and 10.10) the simulation and measurement results of all control strategies are summarized. All of the strategies could be further tuned to enhance their performance. But since the goal is to show an overall concept starting with the modeling and ending with the trajectory tracking control of complex parallel robots the tuning has only performed to some degree.

Controller gains and drive degree of utilization

Comparing the simulation (Tab. 10.9) and measurement results (Tab. 10.10) it can be observed, that especially the position error is in many cases of the same scale, speaking for a good modeling and simulation of the system dynamics. Furthermore the measured η is, except for Comb. ②, always lower than the simulated one. Especially the latter is in the most cases $\bar{\eta} = 99.9796$. This point has to be further discussed.

From the comparison of the results of Comb. ① and ② it can be concluded, that even if in the simulation there is a high drive degree of utilization and a bang-bang behavior this does not necessarily also hold for the experimental results – cp. also Sec. 10.3.5. Two points can be concluded: a drive degree of utilization of $\bar{\eta} = 99.9796$ is reached in the simulation much earlier than at the demonstrator (especially in case of high gains) and higher gains and therefore a faster convergence lead in case of the most here studied control algorithms to a better performance at the demonstrator[3]. Especially in case of the loop shaping approach ⑦, the measured results are one decimal power better than the simulated ones.

The statement about the higher controller gains is further confirmed studying the results of the LQG ⑥(in Sec. 10.4.1) and state-space control ⑩ (in Sec. 10.5.2). These results give a proper (first) hint for the choice of the controller gains in case of the SpiderMill. But beside the size of the gains also their correlation is important. This point can be observed from the comparison of the LQG control (⑥, Tab. 10.10) and the PD-control with LKF (gains C1) ③. In case of the LQG with lower, but better 'harmonized' gains a further improvement of the track-

---

[3]The influence of the chosen observation strategy on η has also to be kept in mind. It can be observed in the simulation and experimental (discussed below) results.

ing performance has been reached, even with a lower $\bar{\eta}$. Also the results for the state-space control ⑩ in Tab. 10.7 indicate, that a better tracking performance is not necessarily related to a higher η. This aspect has already been discussed in Sec. 10.3.5 and is mainly a result of the lower necessary control effort in case of a better trajectory tracking. Consequently, the finally resulting η is a mix between the amplitudes of the actuating variables due to the gains and the reduction of the control effort in case of a good trajectory tracking.

Advanced velocity reconstruction or differentiation

Comparing the measurement results (Tab. 10.10) in case of the model-based control algorithms it is obvious that the strategy for the determination of the velocities has a strong influence on the control performance and consequently also on η. This becomes obvious from the results of the algorithms ①,③, ④ and ⑤, where the same controller gains (C1) have been chosen. The best results are obtained in case of the LKF. Additionally it can be observed that η is significantly larger applying the FOAW algorithms in combination with the model-based control concepts.

In order to shortly discuss the statement in Chap. 9, that for the special actuation principle of the SpiderMill it would also be possible to apply a numerical differentiation of the position signal (spindle length) Tab. 10.8 is given. The measurement results indicate, that in case of the SpiderMill and the defined trajectories, a numerical differentiation can even lead to a slightly better performance than applying the best-fit-FOAW (at least in case of the studied state-space control).

variant	$\overline{\Delta P}$ $[m]$	$\overline{\Delta S}$ $[m/s]$	$\overline{\Delta \dot{S}}$ $[m/s^2]$	$\bar{\eta}$
variant 1, b	$6.1938 \cdot 10^{-3}$	$1.2344 \cdot 10^{-3}$	$2.8204 \cdot 10^{-3}$	40.4107
variant 1, d	$5.8676 \cdot 10^{-3}$	$1.1714 \cdot 10^{-3}$	$2.7482 \cdot 10^{-3}$	40.2329
variant 2, b	$4.2433 \cdot 10^{-3}$	$8.7565 \cdot 10^{-4}$	$2.3421 \cdot 10^{-3}$	40.0803
variant 2, d	$4.1669 \cdot 10^{-3}$	$8.6174 \cdot 10^{-4}$	$2.3394 \cdot 10^{-3}$	40.4098
variant 3, b	$2.8195 \cdot 10^{-3}$	$6.1234 \cdot 10^{-4}$	$2.0083 \cdot 10^{-3}$	40.1781
variant 3, d	$2.7759 \cdot 10^{-3}$	$6.0785 \cdot 10^{-4}$	$2.0413 \cdot 10^{-3}$	40.7326

Table 10.8: State-space control: (b)est-fit-FOAW and numerical (d)ifferentiation

Model-based vs. Non-model-based control

The experimental results show the superior performance in case of the model-based control concepts ① - ⑧ compared to the non-model-based ⑨ - ⑩. The performance of the best non-model-based control strategy, the state-space control ⑩ is still lower than that of the worst model-based approach the PD with gravity compensation ⑧. For the simulation results this does not hold in all cases and is there strongly related to the above discussed early reaching of a high η.[4] But it has to be pointed out, that also the non-model-based concepts provide good results, which are for many applications adequate enough.

In case of the non-model-based strategies the state-space control ⑩ (centralized strategy) has a smaller tracking error than the single-axis PID-control ⑨ (decentralized strategy). This also matches with the theoretical considerations. Due to that, the two not synchronized single axis controller should work against each other to some degree lowering the performance. But this point also depends strongly on the trajectories (or better the paths), which are in this case maybe to symmetrical to enforce this effect very much.

Studying all measurement results it can also be observed, that for well tuned control algorithms (and consequently only not for Comb. 2) the tracking accuracy is always higher in case of trajectories planned for a lower jerk limit (and this in almost all cases with a lower η).

For a concrete task it is always necessary to find a compromise between the trajectory traveling time, tracking accuracy and drive degree of utilization. This holds in particular for the here studied case, where the gains on the traveling time, cp. Tab. 7.2 in case of higher jerk levels are not significant.

Ideal control approach

It is not possible to define an ideal control strategy for the entire class of parallel robots. But the results presented above give some good hints for developing an adequate concept.

In a first step a very simple control strategy like the single axis PID control – requiring no full system model as starting point – should be implemented and the required velocities should be determined based on numerical differentiation (if the encoder resolution is not to low and the level of measurement noise is acceptable). The obtained results for the task related trajectories define a lower bound for the reachable control performance. If the results satisfy the requirements an always more or less costly modeling of the system dynamics can be avoided. If not it should be verified first, if the performance can be enhanced by applying an AW-algorithm or the LKF for the velocity estimation, whereas the latter one should be (at least based on the results above) favored.

One of the most important factors limiting the implementable control concepts and their performances are the available control hardware and the thereby possible sampling rates. Even if advanced strategies like that introduced in Chap. 5

---

[4]It is possible to tune the model-based control algorithms in a way, that also in the simulation their performance is better than that of the non-model-based ones. But here another way has been gone. The tuning has been done on the manipulator and the simulations have been performed with the same controllers and gains (except in case of the reference strategy).

are applied for deriving the dynamic equations of a complex parallel robot, the model complexity always limits the reachable sampling rate. If there is a strong limitation which does not allow to implement a feedback-linearization at least the gravity matrix should be compensated, applying the PD with gravity compensation strategy. If the control requirements are high and if the controller hardware allows to implement it the best choice would be to realize the LQG approach based on the feedback-linearized system. Beside its excellent performance the strategy also avoids the longsome and only heuristically possible tuning process related to other approaches. If the robustness of the control becomes an even more important issue the loop shaping approach should be favored.

10.7 Conclusions and future works

In this section one reference tracking control strategy based on the feedback-linearized (and stabilized) dynamics as well as extensions of it has been discussed in detail. Furthermore additional standard model- and non-model-based approaches have been analyzed. As expected the model-based concepts show a superior control performance compared to the non-model-based ones. The performances of all concepts can be further improved by a tuning process. Beside an overall tuning also a tuning of the parameters with respect to the individual trajectories would be a possible extension of the strategy. These, in the most cases heuristic optimization steps have only to some degree performed in this thesis, since its overall goal is the development of a holistic modeling and trajectory tracking control strategy for complex parallel robots.

As a consequence of the results regarding the controller gains and the drive degree of utilization it seems reasonable to further extend the simulation model by the entire drive chains including the ball-screw-systems. Additionally a consideration of the friction influences will allow a better tuning of the controllers.

Several extensions in case of the trajectory tracking are possible. Direct extensions would be the realization of (further) robust and adaptive control concepts. Thereby, especially the adaptive ones would lead to better handling of model uncertainties or changing (e.g. payloads, friction parameters) dynamics by directly improving the controller parameters. In the same direction also the concept of iterative learning control works, which is particularly applicable to the here considered cyclic processes.

The table columns are grouped as ΔP_i, ΔS_i, $\Delta \dot S_i$, η_i for $i = 1 \dots 7$, and a final mean column $\overline{\Delta P}\ [m]$, $\overline{\Delta S}\ [m/s]$, $\overline{\Delta \dot S}\ [m/s^2]$, $\overline{\eta}$.

Method	Qty	$i=1$	$i=2$	$i=3$	$i=4$	$i=5$	$i=6$	$i=7$	mean
① Comb. 1	ΔP	$1.3570\cdot10^{-4}$	$1.3340\cdot10^{-4}$	$1.3679\cdot10^{-4}$	$1.3264\cdot10^{-4}$	$1.3178\cdot10^{-4}$	$1.5194\cdot10^{-4}$	$1.3832\cdot10^{-4}$	$1.3723\cdot10^{-4}$
	ΔS	$2.5473\cdot10^{-5}$	$2.5321\cdot10^{-5}$	$2.6054\cdot10^{-5}$	$2.5885\cdot10^{-5}$	$2.6446\cdot10^{-5}$	$2.9158\cdot10^{-5}$	$2.2233\cdot10^{-5}$	$2.5796\cdot10^{-5}$
	$\Delta \dot S$	$2.8479\cdot10^{-3}$	$2.8710\cdot10^{-3}$	$2.9136\cdot10^{-3}$	$2.8748\cdot10^{-3}$	$2.8927\cdot10^{-3}$	$3.0837\cdot10^{-3}$	$2.6581\cdot10^{-3}$	$2.8774\cdot10^{-3}$
	η	89.2536	90.0400	89.1789	89.1252	89.2320	89.2383	88.6290	89.2424
② Comb. 2	ΔP	$1.8926\cdot10^{-5}$	$2.4858\cdot10^{-5}$	$2.6151\cdot10^{-5}$	$1.3479\cdot10^{-5}$	$1.4586\cdot10^{-5}$	$6.2375\cdot10^{-5}$	$2.3832\cdot10^{-5}$	$2.6315\cdot10^{-5}$
	ΔS	$3.2081\cdot10^{-6}$	$4.3147\cdot10^{-6}$	$4.5957\cdot10^{-6}$	$2.1099\cdot10^{-6}$	$2.4396\cdot10^{-6}$	$1.3420\cdot10^{-5}$	$4.1017\cdot10^{-6}$	$4.8842\cdot10^{-6}$
	$\Delta \dot S$	$1.3759\cdot10^{-4}$	$1.7785\cdot10^{-4}$	$1.9304\cdot10^{-4}$	$1.0373\cdot10^{-4}$	$1.2221\cdot10^{-4}$	$3.5977\cdot10^{-4}$	$1.7219\cdot10^{-4}$	$1.8091\cdot10^{-4}$
	η	31.8607	32.4666	32.6578	32.7766	32.4749	33.5687	34.3075	32.8732
③ PD with LKF	ΔP	$2.5832\cdot10^{-4}$	$2.8426\cdot10^{-4}$	$2.9573\cdot10^{-4}$	$3.0383\cdot10^{-4}$	$3.2892\cdot10^{-4}$	$3.7054\cdot10^{-4}$	$3.2120\cdot10^{-4}$	$3.0897\cdot10^{-4}$
	ΔS	$4.6024\cdot10^{-5}$	$5.1430\cdot10^{-5}$	$5.2607\cdot10^{-5}$	$5.3430\cdot10^{-5}$	$5.9155\cdot10^{-5}$	$6.7117\cdot10^{-5}$	$5.0101\cdot10^{-5}$	$5.4266\cdot10^{-5}$
	$\Delta \dot S$	$4.8377\cdot10^{-3}$	$5.0589\cdot10^{-3}$	$5.1262\cdot10^{-3}$	$5.2053\cdot10^{-3}$	$5.4979\cdot10^{-3}$	$5.8653\cdot10^{-3}$	$4.4847\cdot10^{-3}$	$5.2109\cdot10^{-3}$
	η	99.9848	99.9839	99.9838	99.9803	99.9767	99.9756	99.9720	99.9796
④ PD with end-fit-FOAW	ΔP	$1.9004\cdot10^{-2}$	$1.0794\cdot10^{-2}$	$1.2521\cdot10^{-2}$	$1.1316\cdot10^{-2}$	$1.4878\cdot10^{-2}$	$1.2977\cdot10^{-2}$	$1.0600\cdot10^{-2}$	$1.3116\cdot10^{-2}$
	ΔS	$3.1370\cdot10^{-3}$	$1.8567\cdot10^{-3}$	$2.1379\cdot10^{-3}$	$2.0816\cdot10^{-3}$	$2.7743\cdot10^{-3}$	$2.4594\cdot10^{-3}$	$1.6340\cdot10^{-3}$	$2.2973\cdot10^{-3}$
	$\Delta \dot S$	$7.0196\cdot10^{-2}$	$5.7291\cdot10^{-2}$	$6.1810\cdot10^{-2}$	$6.2277\cdot10^{-2}$	$7.1592\cdot10^{-2}$	$6.4070\cdot10^{-2}$	$5.0330\cdot10^{-2}$	$6.2509\cdot10^{-2}$
	η	99.9848	99.9839	99.9838	99.9803	99.9767	99.9756	99.9720	99.9796
⑤ PD best-fit-FOAW	ΔP	$3.7079\cdot10^{-4}$	$5.0885\cdot10^{-4}$	$3.5916\cdot10^{-4}$	$3.4616\cdot10^{-4}$	$9.0191\cdot10^{-4}$	$3.5004\cdot10^{-4}$	$4.4686\cdot10^{-4}$	$4.6911\cdot10^{-4}$
	ΔS	$6.4419\cdot10^{-5}$	$8.7518\cdot10^{-5}$	$6.2682\cdot10^{-5}$	$5.9602\cdot10^{-5}$	$1.4172\cdot10^{-4}$	$6.1583\cdot10^{-5}$	$7.0854\cdot10^{-5}$	$7.8339\cdot10^{-5}$
	$\Delta \dot S$	$8.2123\cdot10^{-3}$	$9.2892\cdot10^{-3}$	$8.3515\cdot10^{-3}$	$8.1278\cdot10^{-3}$	$1.1278\cdot10^{-2}$	$8.3397\cdot10^{-3}$	$8.1974\cdot10^{-3}$	$8.8280\cdot10^{-3}$
	η	99.9848	99.9839	99.9838	99.9803	99.9767	99.9756	99.9720	99.9796
⑥ LQG	ΔP	$2.1433\cdot10^{-4}$	$2.2970\cdot10^{-4}$	$2.2642\cdot10^{-4}$	$2.1426\cdot10^{-4}$	$2.3169\cdot10^{-4}$	$2.3622\cdot10^{-4}$	$2.5797\cdot10^{-4}$	$2.3009\cdot10^{-4}$
	ΔS	$3.7899\cdot10^{-5}$	$4.0883\cdot10^{-5}$	$4.0762\cdot10^{-5}$	$3.8486\cdot10^{-5}$	$4.1960\cdot10^{-5}$	$4.2804\cdot10^{-5}$	$4.0889\cdot10^{-5}$	$4.0526\cdot10^{-5}$
	$\Delta \dot S$	$5.4888\cdot10^{-3}$	$5.7022\cdot10^{-3}$	$5.7617\cdot10^{-3}$	$5.5913\cdot10^{-3}$	$5.8593\cdot10^{-3}$	$5.9182\cdot10^{-3}$	$5.3626\cdot10^{-3}$	$5.6692\cdot10^{-3}$
	η	99.9848	99.9839	99.9838	99.9803	99.9767	99.9756	99.9720	99.9796
⑦ Loop Shaping	ΔP	$1.1393\cdot10^{-3}$	$1.1801\cdot10^{-3}$	$1.1593\cdot10^{-3}$	$1.1676\cdot10^{-3}$	$1.1728\cdot10^{-3}$	$1.1673\cdot10^{-3}$	$1.2778\cdot10^{-3}$	$1.1806\cdot10^{-3}$
	ΔS	$2.1244\cdot10^{-4}$	$2.1936\cdot10^{-4}$	$2.2131\cdot10^{-4}$	$2.2351\cdot10^{-4}$	$2.4050\cdot10^{-4}$	$2.3787\cdot10^{-4}$	$2.0601\cdot10^{-4}$	$2.2300\cdot10^{-4}$
	$\Delta \dot S$	$5.4812\cdot10^{-3}$	$4.9307\cdot10^{-3}$	$4.9449\cdot10^{-3}$	$4.9200\cdot10^{-3}$	$5.4807\cdot10^{-3}$	$5.5841\cdot10^{-3}$	$5.0245\cdot10^{-3}$	$5.0666\cdot10^{-3}$
	η	99.9848	99.9839	99.9838	99.9803	99.9767	99.9756	99.9720	99.9796
⑧ PD with gravity compensation	ΔP	$1.0192\cdot10^{-3}$	$1.0024\cdot10^{-3}$	$1.0403\cdot10^{-3}$	$1.0458\cdot10^{-3}$	$1.0576\cdot10^{-3}$	$1.0760\cdot10^{-3}$	$1.1734\cdot10^{-3}$	$1.0592\cdot10^{-3}$
	ΔS	$1.7385\cdot10^{-4}$	$1.7416\cdot10^{-4}$	$1.8095\cdot10^{-4}$	$1.7640\cdot10^{-4}$	$1.8186\cdot10^{-4}$	$1.9105\cdot10^{-4}$	$1.8610\cdot10^{-4}$	$1.8063\cdot10^{-4}$
	$\Delta \dot S$	$1.5895\cdot10^{-2}$	$1.7211\cdot10^{-2}$	$1.6599\cdot10^{-2}$	$1.6894\cdot10^{-2}$	$1.6676\cdot10^{-2}$	$1.7740\cdot10^{-2}$	$1.6879\cdot10^{-2}$	$1.6842\cdot10^{-2}$
	η	99.9848	99.9839	99.9838	99.9803	99.9767	99.9756	99.9720	99.9796
⑨ Single axis PID	ΔP	$1.6134\cdot10^{-3}$	$1.5428\cdot10^{-3}$	$1.5347\cdot10^{-3}$	$1.6803\cdot10^{-3}$	$1.6622\cdot10^{-3}$	$1.9421\cdot10^{-3}$	$1.8875\cdot10^{-3}$	$1.6947\cdot10^{-3}$
	ΔS	$2.7095\cdot10^{-3}$	$2.5740\cdot10^{-4}$	$2.5941\cdot10^{-4}$	$2.4871\cdot10^{-4}$	$2.4949\cdot10^{-4}$	$2.9829\cdot10^{-4}$	$3.0563\cdot10^{-4}$	$2.6998\cdot10^{-4}$
	$\Delta \dot S$	$5.7207\cdot10^{-3}$	$5.7873\cdot10^{-3}$	$5.8074\cdot10^{-3}$	$5.8108\cdot10^{-3}$	$5.8318\cdot10^{-3}$	$5.8858\cdot10^{-3}$	$5.3786\cdot10^{-3}$	$5.7461\cdot10^{-3}$
	η	99.9848	99.9839	99.9838	99.9803	99.9767	99.9756	99.9720	99.9796
⑩ State-space control	ΔP	$4.5649\cdot10^{-4}$	$5.3045\cdot10^{-4}$	$4.9697\cdot10^{-4}$	$5.7920\cdot10^{-4}$	$5.0368\cdot10^{-4}$	$6.3890\cdot10^{-4}$	$5.3454\cdot10^{-4}$	$5.3432\cdot10^{-4}$
	ΔS	$9.1143\cdot10^{-5}$	$1.0743\cdot10^{-4}$	$9.9342\cdot10^{-5}$	$1.1438\cdot10^{-4}$	$9.8517\cdot10^{-5}$	$1.2673\cdot10^{-4}$	$8.7329\cdot10^{-5}$	$1.0355\cdot10^{-4}$
	$\Delta \dot S$	$5.2722\cdot10^{-3}$	$5.4142\cdot10^{-3}$	$5.3468\cdot10^{-3}$	$5.2921\cdot10^{-3}$	$5.4182\cdot10^{-3}$	$5.5770\cdot10^{-3}$	$4.8448\cdot10^{-3}$	$5.3093\cdot10^{-3}$
	η	99.9848	99.9839	99.9838	99.9803	99.9767	99.9756	99.9720	99.9796

Table 10.9: Simulation results

Method	Metric	1	2	3	4	5	6	7	Total
① Comb. 1	ΔP	$1.0712\cdot10^{-5}$	$1.1377\cdot10^{-5}$	$1.3144\cdot10^{-4}$	$1.1379\cdot10^{-4}$	$1.4221\cdot10^{-4}$	$2.5967\cdot10^{-4}$	$1.3868\cdot10^{-4}$	$1.4381\cdot10^{-4}$
	ΔS	$6.9446\cdot10^{-5}$	$1.3113\cdot10^{-4}$	$1.3439\cdot10^{-4}$	$1.1303\cdot10^{-4}$	$1.3455\cdot10^{-4}$	$9.8342\cdot10^{-5}$	$1.3094\cdot10^{-4}$	$1.1597\cdot10^{-4}$
	$\Delta \dot{S}$	$1.5017\cdot10^{-3}$	$1.5903\cdot10^{-3}$	$1.8018\cdot10^{-3}$	$1.6233\cdot10^{-3}$	$1.9866\cdot10^{-3}$	$3.6136\cdot10^{-3}$	$1.8719\cdot10^{-3}$	$1.9984\cdot10^{-3}$
	η	45.1247	48.2282	48.2199	41.6645	48.1506	53.1748	55.4417	48.5721
② Comb. 2	ΔP	$1.4064\cdot10^{-4}$	$1.3932\cdot10^{-4}$	$1.4006\cdot10^{-4}$	$1.4449\cdot10^{-3}$	$1.6192\cdot10^{-3}$	$1.6876\cdot10^{-3}$	$1.4897\cdot10^{-3}$	$1.4916\cdot10^{-3}$
	ΔS	$2.6835\cdot10^{-4}$	$3.2718\cdot10^{-4}$	$3.3090\cdot10^{-4}$	$3.2944\cdot10^{-4}$	$3.8786\cdot10^{-4}$	$3.4330\cdot10^{-4}$	$3.4682\cdot10^{-4}$	$3.3341\cdot10^{-4}$
	$\Delta \dot{S}$	$5.0673\cdot10^{-3}$	$5.1542\cdot10^{-3}$	$5.2191\cdot10^{-3}$	$5.3646\cdot10^{-3}$	$6.3396\cdot10^{-3}$	$6.8749\cdot10^{-3}$	$5.4938\cdot10^{-3}$	$5.6448\cdot10^{-3}$
	η	34.0060	34.7442	34.7550	31.2114	34.4052	35.6712	39.2438	34.8624
③ PD with LKF	ΔP	$6.7648\cdot10^{-5}$	$7.0618\cdot10^{-5}$	$8.3015\cdot10^{-5}$	$6.9673\cdot10^{-5}$	$7.8514\cdot10^{-5}$	$1.0714\cdot10^{-4}$	$8.8391\cdot10^{-5}$	$8.0714\cdot10^{-5}$
	ΔS	$6.1783\cdot10^{-5}$	$1.2294\cdot10^{-4}$	$1.2557\cdot10^{-4}$	$1.0431\cdot10^{-4}$	$1.2288\cdot10^{-4}$	$7.5536\cdot10^{-5}$	$1.2227\cdot10^{-4}$	$1.0504\cdot10^{-4}$
	$\Delta \dot{S}$	$1.7240\cdot10^{-3}$	$1.6546\cdot10^{-3}$	$1.7581\cdot10^{-3}$	$1.7553\cdot10^{-3}$	$1.6326\cdot10^{-3}$	$1.9859\cdot10^{-3}$	$1.8798\cdot10^{-3}$	$1.7701\cdot10^{-3}$
	η	68.7163	69.7368	70.4216	64.1555	64.1177	66.1023	75.9476	68.4568
④ PD with end-fit-FOAW	ΔP	$6.3367\cdot10^{-5}$	$6.6419\cdot10^{-5}$	$6.8840\cdot10^{-5}$	$6.3650\cdot10^{-5}$	$6.7037\cdot10^{-5}$	$7.6186\cdot10^{-5}$	$7.4903\cdot10^{-5}$	$6.8629\cdot10^{-5}$
	ΔS	$1.3642\cdot10^{-4}$	$1.8938\cdot10^{-4}$	$1.9370\cdot10^{-4}$	$1.6885\cdot10^{-4}$	$1.8925\cdot10^{-4}$	$1.6758\cdot10^{-4}$	$1.9321\cdot10^{-4}$	$1.7691\cdot10^{-4}$
	$\Delta \dot{S}$	$1.3157\cdot10^{-2}$	$1.3220\cdot10^{-2}$	$1.3382\cdot10^{-2}$	$1.2546\cdot10^{-2}$	$1.2796\cdot10^{-2}$	$1.3412\cdot10^{-2}$	$1.3159\cdot10^{-2}$	$1.3096\cdot10^{-2}$
	η	91.9753	92.2497	92.3206	91.3076	91.3652	91.0892	92.3089	91.8024
⑤ PD with best-fit-FOAW	ΔP	$1.6395\cdot10^{-5}$	$1.6999\cdot10^{-4}$	$1.8216\cdot10^{-4}$	$1.7525\cdot10^{-4}$	$1.8249\cdot10^{-4}$	$2.1268\cdot10^{-4}$	$1.9767\cdot10^{-4}$	$1.8346\cdot10^{-4}$
	ΔS	$7.1856\cdot10^{-5}$	$1.3221\cdot10^{-4}$	$1.3494\cdot10^{-4}$	$1.1442\cdot10^{-4}$	$1.3301\cdot10^{-4}$	$8.5772\cdot10^{-5}$	$1.3463\cdot10^{-4}$	$1.1526\cdot10^{-4}$
	$\Delta \dot{S}$	$6.5188\cdot10^{-3}$	$6.5407\cdot10^{-3}$	$6.6120\cdot10^{-3}$	$6.6522\cdot10^{-3}$	$6.5978\cdot10^{-3}$	$6.7962\cdot10^{-3}$	$6.6136\cdot10^{-3}$	$6.6045\cdot10^{-3}$
	η	90.2156	91.2759	91.0585	89.3107	88.5960	88.5154	91.5482	90.0743
⑥ LQG	ΔP	$6.3937\cdot10^{-4}$	$6.8148\cdot10^{-4}$	$8.8011\cdot10^{-4}$	$6.3149\cdot10^{-4}$	$7.4164\cdot10^{-4}$	$1.0193\cdot10^{-5}$	$8.1341\cdot10^{-4}$	$7.7241\cdot10^{-4}$
	ΔS	$6.2064\cdot10^{-5}$	$1.2339\cdot10^{-4}$	$1.2674\cdot10^{-4}$	$1.0398\cdot10^{-4}$	$1.2284\cdot10^{-4}$	$7.3992\cdot10^{-5}$	$1.2230\cdot10^{-4}$	$1.0504\cdot10^{-4}$
	$\Delta \dot{S}$	$1.6239\cdot10^{-3}$	$1.6075\cdot10^{-3}$	$1.8082\cdot10^{-3}$	$1.6697\cdot10^{-3}$	$1.7143\cdot10^{-3}$	$2.0667\cdot10^{-3}$	$2.0598\cdot10^{-3}$	$1.7929\cdot10^{-3}$
	η	57.9516	59.1421	59.2397	53.9566	56.1659	57.2900	67.6886	58.7764
⑦ Loop Shaping	ΔP	$1.7404\cdot10^{-4}$	$1.9102\cdot10^{-4}$	$2.1632\cdot10^{-4}$	$1.6769\cdot10^{-4}$	$2.0138\cdot10^{-4}$	$2.3367\cdot10^{-4}$	$2.2104\cdot10^{-4}$	$2.0074\cdot10^{-4}$
	ΔS	$7.8597\cdot10^{-5}$	$1.1426\cdot10^{-4}$	$1.4581\cdot10^{-4}$	$1.2100\cdot10^{-4}$	$1.4563\cdot10^{-4}$	$1.0026\cdot10^{-4}$	$1.4438\cdot10^{-4}$	$1.2528\cdot10^{-4}$
	$\Delta \dot{S}$	$9.0968\cdot10^{-4}$	$1.2096\cdot10^{-3}$	$1.3470\cdot10^{-3}$	$9.8108\cdot10^{-4}$	$1.3069\cdot10^{-3}$	$1.6083\cdot10^{-3}$	$1.4841\cdot10^{-3}$	$1.2638\cdot10^{-3}$
	η	77.6041	80.6665	79.9388	68.8296	72.804	74.6095	82.3371	76.6894
⑧ PD with gravity compensation	ΔP	$1.3734\cdot10^{-3}$	$1.6329\cdot10^{-3}$	$1.6975\cdot10^{-3}$	$1.3568\cdot10^{-3}$	$2.0881\cdot10^{-3}$	$2.8741\cdot10^{-3}$	$1.9975\cdot10^{-3}$	$1.8600\cdot10^{-3}$
	ΔS	$3.5720\cdot10^{-4}$	$4.0239\cdot10^{-4}$	$4.1497\cdot10^{-4}$	$3.4155\cdot10^{-4}$	$4.6095\cdot10^{-4}$	$5.5983\cdot10^{-4}$	$4.3631\cdot10^{-4}$	$4.2474\cdot10^{-4}$
	$\Delta \dot{S}$	$1.7933\cdot10^{-3}$	$5.5035\cdot10^{-3}$	$6.0632\cdot10^{-3}$	$1.5662\cdot10^{-3}$	$9.8625\cdot10^{-3}$	$1.5270\cdot10^{-3}$	$9.5655\cdot10^{-3}$	$7.0892\cdot10^{-3}$
	η	38.2664	45.0004	45.3105	35.9426	50.4209	62.6686	56.1165	47.6751
⑨ Single axis PID	ΔP	$2.6640\cdot10^{-3}$	$2.7384\cdot10^{-3}$	$3.9766\cdot10^{-3}$	$2.5675\cdot10^{-3}$	$2.8372\cdot10^{-3}$	$5.7850\cdot10^{-3}$	$4.4633\cdot10^{-3}$	$3.5760\cdot10^{-3}$
	ΔS	$5.1266\cdot10^{-4}$	$5.8770\cdot10^{-4}$	$8.5329\cdot10^{-4}$	$5.7762\cdot10^{-4}$	$6.3470\cdot10^{-4}$	$9.8128\cdot10^{-4}$	$8.3276\cdot10^{-4}$	$7.1143\cdot10^{-4}$
	$\Delta \dot{S}$	$1.5063\cdot10^{-3}$	$1.8162\cdot10^{-3}$	$1.9601\cdot10^{-3}$	$1.2902\cdot10^{-3}$	$1.7027\cdot10^{-3}$	$2.2620\cdot10^{-3}$	$2.2606\cdot10^{-3}$	$1.8283\cdot10^{-3}$
	η	38.2387	39.4290	39.8878	36.0287	38.9373	41.3505	45.0380	39.8443
⑩ State-space control	ΔP	$2.6022\cdot10^{-3}$	$2.6499\cdot10^{-3}$	$2.6748\cdot10^{-3}$	$2.8778\cdot10^{-3}$	$3.0918\cdot10^{-3}$	$3.2597\cdot10^{-3}$	$2.5805\cdot10^{-3}$	$2.8195\cdot10^{-3}$
	ΔS	$5.3401\cdot10^{-4}$	$6.1107\cdot10^{-4}$	$6.1788\cdot10^{-4}$	$6.0055\cdot10^{-4}$	$6.7202\cdot10^{-4}$	$6.5319\cdot10^{-4}$	$5.9764\cdot10^{-4}$	$6.1234\cdot10^{-4}$
	$\Delta \dot{S}$	$1.5074\cdot10^{-3}$	$2.0104\cdot10^{-3}$	$2.0998\cdot10^{-3}$	$1.3774\cdot10^{-3}$	$2.1506\cdot10^{-3}$	$2.4003\cdot10^{-3}$	$2.5120\cdot10^{-3}$	$2.0083\cdot10^{-3}$
	η	38.8109	39.8621	40.1099	36.8314	39.4753	42.1826	43.9743	40.1781

Column header units — index columns: ΔP_i, ΔS_i, $\Delta \dot{S}_i$, η_i; total column: $\Delta P\,[m]$, $\Delta S\,[m/s]$, $\Delta \dot{S}\,[m/s^2]$, $\bar{\eta}$.

Table 10.10: Measurement results

11 Conclusions and Future Work

In this thesis a holistic concept for model-based trajectory tracking control of complex parallel robots has been introduced and demonstrated at the SpiderMill.

After introducing the planar parallel robot SpiderMill and performing a structural analysis including its DOFs, workspace, singularities as well as its direct and inverse kinematics, a new strategy for modeling of parallel robots with redundancies has been introduced and applied to the SpiderMill. Based on the simplified structure of the robot several standard strategies for dynamic modeling have been used in order to derive analytical dynamic models. The quality of the obtained equations of motion justify the simplification strategy. For one of the models all further steps in terms of model-based trajectory tracking control have been performed, starting with the identification of its rigid body and friction parameters, to further improve the model quality.

Since in literature, only a few trajectory planning strategies – with several disadvantages in case of the class of manipulators studied here – can be found, a new approach for the planning of path-constrained smooth trajectories (PCSMTOPTs) has been introduced in the next step.

Based on the feedback linearization of the non-linear system model it has been possible to apply linear control theory for the design of trajectory tracking controllers. Since already the feedback linearization – and also most control strategies – requires the spindle velocities, first different strategies for the velocity reconstruction had to be analyzed.

Finally one model-based trajectory tracking control strategy in combination with the introduced observation algorithms has been analyzed in detail. Additionally, further common model- and non-model-based control concepts have been implemented and tested in a simulation environment as well as at the physical demonstrator.

Since all chapters have been written in form of closed descriptions with a direct discussion of future work, the individual aspects are not recapitulated again. The next step on the way to a rapid prototyping process would be the extension of the pure position control to a hybrid control with all related challenges. Afterwards also deflection and oscillation effects should be studied based on dynamic models considering flexible links and handled in the tracking and force control algorithms.

A Kinematic, dynamic and friction parameters

In Tab. A.1 the kinematic and in Tab. A.2 the dynamic paramters of the Spider-Mill taken from the MSC.ADAMS (and SimMechanics) model of the SpiderMill are summarized. The other table is a result of the parameter identification in Chap. 6. Assuming, that the masses m_1, m_2 and m_{CM} are known and applying the relationships in Eq.(6.31), the physical parameters can be calculated from the identified ones as summarized in Tab. A.3. The parameters in column two are used analyzing model based control strategies with the SimMechanics model of the robot whereas that of the last column (as well as the identified friction parameters) are used for implementing the control strategies at the demonstrator.

MSC.ADAMS / SimMechanics parameters					
a_1	220.0000 mm	d_A	365.6929 mm	α	48.3202°
a_{21}	68.0000 mm	d_s	400.0000 mm	β	34.2911°
a_{22}	78.0000 mm	d_0	2000.0000 mm	ϵ	3.7270°
a_3	210.0000 mm	d_1	1130.0000 mm	θ_{off}	3.5277°
a_A	184.0000 mm	d_2	184.2830 mm	α_1	28.8700°
b_1	54.5422 mm	d_3	1030.0000 mm	α_2	66.9520°
b_3	59.6359 mm	d_4	128.0000 mm	α_3	79.5511°
δ_1	203.4886 mm	d_5	256.0000 mm	α_4	30.0563°
δ_2	106.7721 mm	d_{2li}	153.5176 mm	α_5	68.1608°
δ_3	111.6531 mm	d_{2lo}	319.1984 mm	α_6	38.8877°
δ_4	219.2298 mm	d_{2ri}	103.4795 mm	μ_{2li}	63.7080°
δ_5	137.8966 mm	d_{2ro}	407.2935 mm	μ_{2lo}	25.5432°
δ_6	203.8881 mm			μ_{2ri}	48.9182°
				μ_{2ro}	45.0000°

Table A.1: Kinematic parameters

MSC.ADAMS / SimMechanics parameters					
m_1	50.9598 kg	I_{m12}	6.3195 kg m^2	d_{m1}	564.9293 mm
m_2	7.9538 kg	I_{m32}	1.6173 kg m^2	d_{m2}	99.0719 mm
m_3	14.5773 kg	I_{m1i2}	4.5528 kg m^2	d_{m3}	515.0000 mm
m_{CM}	17.8900 kg	I_{m1o2}	1.7667 kg m^2	d_{m4}	128.2713 mm
m_{1i}	38.3259 kg	I_{m3i2}	1.1704 kg m^2	g	9.80665 m s^{-2}
m_{1o}	12.6339 kg	I_{m3o2}	0.4474 kg m^2		
m_{3i}	10.4404 kg				
m_{3o}	4.1397 kg				

Table A.2: Dynamic parameters (nominal values)

parameter	MSC.ADAMS / SimMechanics	estimation MSC.ADAMS	estimation demonstrator
m_3	14.5773 kg	7.0263 kg	−2.1649 kg
d_{m1}	0.5649 kg	0.7357 m	0.0368 m
d_{m3}	0.5150 kg	1.0513 m	1.5721 m
I_{m12}	6.3195 kgm^2	4.8961 kg m^2	42.0428 kg m^2
I_{m32}	1.6173 kg m^2	−2.5073 kg m^2	−4.4983 kg m^2
r_{al}	−	−	0.0520
r_{ar}	−	−	0.0395
r_{pal}	−	−	−23.6001
r_{par}	−	−	−23.9920
r_{pll}	−	−	−9.3241
r_{plr}	−	−	−5.2453
r_{pM}	−	−	−2.6301

Table A.3: Estimated dynamic and friction parameters

B Concepts of dynamic modeling

This appendix is given in order to summarize the theoretical background of the different approaches for dynamic modeling used in this thesis in a compact form. None of the here described concepts has been developed by the author, they are all well known in literature and only recapitulated below. Their theory and mathematical descriptions are mainly taken from [Tsa99, SHV06, SSVO09, AGH05].

B.1 Constraints and generalized coordinates[1]

In a mechanical system several types of constraints exist. According to [Tsa99] a kinematic constraint imposes some conditions on the relative motion between a pair of bodies. The most frequently constraints in robotics are those provided by the joints of a manipulator. All constraints can be divided into two classes. A kinematic constraint is holonomic if its conditions can be expressed as algebraic equations of their coordinates and the time in the following form

$$f(\mathbf{x}_1, \mathbf{x}_2, \ldots, t) = 0 \tag{B.1}$$

where $\mathbf{x}_i$ denotes the coordinates of a rigid body (or particle), cp. Eq. (3.4) and (3.6). All other constraints are nonholonomic.

Applying the Lagrangian approach, cp. Sec. B.2, two kinds of formulations of the equations of motion have to be distinguished [Tsa99]. In case of the second type the equations are formulated in terms of a minimal set of independent generalized coordinates (number: N_q) and generalized forces. Using this formulation all forces of constraints in the joints do not appear in the final equations, whose number is

$$N_E = N_q = F. \tag{B.2}$$

The resulting system of differential equations is in general more complicated than the equations in case of the Lagrangian of the first type.

In many cases, e.g. for parallel robots, it is easier to describe a robot by more co-ordinates $N_q > F$. For that case the coordinates are not independent anymore and

---

[1]This section is a recapitulation of the results in [Tsa99].

an appropriate set of constraint equations N_C is required for solving the equations of motion using e.g. the method of Lagrangian multipliers

$$N_E = N_q = F + N_C. \tag{B.3}$$

The resulting equations N_E are called Lagrangian equations of the first type and are defining as set of DAEs. They are containing the unknown forces of constraint as the Lagrange multipliers, which are determined in an intermediate step, while calculating the generalized actuator forces. In general, the equations are compact. The approach of the first type is applicable to mechanical systems with either holonomic or nonholonomic constraints. Furthermore, it is more suitable for modeling parallel than serial robots, since the first ones have numerous kinematic constraints due to the presence of the closed loops. In contrast the approach of the second type is only applicable to systems with holonomic constraints and is particularly suitable for serial robots [Tsa99].

B.2 Lagrangian equations

B.2.1 Lagrangian equations of the second type

With the Lagrange approach the equations of motion can be derived in a systematic way independently of the chosen reference coordinate frame. The Lagrangian $\mathcal{L}$ of a mechanical system is a function of the generalized coordinates

$$\mathcal{L}(\mathbf{q}, \dot{\mathbf{q}}) = \mathcal{T}(\mathbf{q}, \dot{\mathbf{q}}) - \mathcal{U}(\mathbf{q}) \tag{B.4}$$

where $\mathcal{T}$ is the kinetic and $\mathcal{U}$ the potential energy of the system, their detailed derivation can be found in [Tsa99]. The individual Lagrange equations can be expressed by

$$\frac{d}{dt}\left(\frac{\partial \mathcal{L}}{\partial \dot{q}_i}\right) - \frac{\partial \mathcal{L}}{\partial q_i} = \boldsymbol{\tau}_i, \quad i = 1, \dots, N_q = N_E \tag{B.5}$$

where the $\boldsymbol{\tau}_i$ are the generalized forces associated to the generalized coordinates q_i, or in compact form

$$\frac{d}{dt}\left(\frac{\partial \mathcal{L}}{\partial \dot{\mathbf{q}}}\right) - \frac{\partial \mathcal{L}}{\partial \mathbf{q}} = \boldsymbol{\tau}. \tag{B.6}$$

In case of serial robots the generalized coordinates $\mathbf{q}$ correspond to the joint variables. Three groups of nonconservative forces contribute to the generalized forces:

the joint actuator torques[2], the joint friction torques as well as the torques induced by the end-effector forces due to a contact with the environment [SSVO09], cp. B.5.

B.2.2 Lagrangian equations of the first type

The Lagrangian equations of the first type use redundant coordinates. To solve the system a set of algebraic constraint equations derived from the kinematics of the mechanisms is required. The equations and their derivatives have to be adjoined to the equations of motion to produce a system with as many equations as unknowns. The Lagrangian equations of the first type – neglecting friction – are defined by [Tsa99] in the following form

$$\frac{d}{dt}\left(\frac{\partial \mathcal{L}}{\partial \dot{q}_i}\right) - \frac{\partial \mathcal{L}}{\partial q_i} = \boldsymbol{\tau}_i + \sum_{j=1}^{N_C} \lambda_j \frac{\partial \Gamma_j}{\partial q_i}, \quad i = 1,\ldots,N_q \qquad (B.7)$$

where Γ_j describes the jth constraint function, N_C is the number of the constraint functions and the λ_js are the Lagrangian multipliers (corresponding to the generalized reaction/constraining forces in the joints). The number of coordinates N_q can be calculated by Eq. (B.3). For solving the equations of motion they are split into two sets. One contains the Lagrange multipliers as only unknowns the other one additionally the actuator torques [Tsa99]. The first N_C equations are associated with the redundant coordinates and the remaining $N_E - N_C$ with the actuated joints. The first set can be written as

$$\sum_{j=1}^{N_C} \lambda_j \frac{\partial \Gamma_j}{\partial q_i} = \frac{d}{dt}\left(\frac{\partial \mathcal{L}}{\partial \dot{q}_i}\right) - \frac{\partial \mathcal{L}}{\partial q_i} - \boldsymbol{\tau}_i^{ext}, \quad i = 1,\ldots,N_C. \qquad (B.8)$$

The $\boldsymbol{\tau}_i^{ext}$ represents the torques contributed by an externally applied torque (also friction terms can be modeled as part of it), if such a torque exists. In case of the IDP the $\boldsymbol{\tau}_i^{ext}$s are given and therefore the right side of Eq. (B.8) is known. By writing Eq. (B.8) once for each redundant coordinate yields a system of N_C linear equations, from which the N_C Lagrangian multipliers can be calculated. Once they are found the actuator torques can be calculated from the remaining equations (second set), which can be written as

$$\boldsymbol{\tau}_i = \frac{d}{dt}\left(\frac{\partial \mathcal{L}}{\partial \dot{q}_i}\right) - \frac{\partial \mathcal{L}}{\partial q_i} - \sum_{j=1}^{N_C} \lambda_j \frac{\partial \Gamma_j}{\partial q_i}, \quad i = N_C + 1,\ldots,N_q \qquad (B.9)$$

where the $\boldsymbol{\tau}_i$s are the actuator torques.

---

[2]If not explicitly defined in another way, the term *torque* is used as a synonym for the generalized forces, which can be torques or linear forces.

B.3 Newton-Euler approach

For serial manipulators the Newton-Euler and the Lagrangian approaches lead to exactly the same final models [SHV06], which means that its algebraic equations of motion can be rearranged, that they are identical. In case of parallel robots the finally calculated torques match, but the individual terms are often combined in a different way. In the most cases the individual summands cannot be recombined in a manner, that identical algebraic equations are obtained. This does also hold for all other approaches.

Whereas in case of the Lagrangian approaches the robot is treated as a whole, in contrast the Newton-Euler approach treats every link of a robot separately. The approach incorporates all forces and moments acting on an individual link. Therefore in a first step all links are cut free and their balances of forces and moments are set up. Over the generalized forces (coupling forces and torques) at the joints the individual sets of equations are related to each other. As a consequence of the strategy the forces of constraints between two adjacent links are included in the final dynamic equations. This information can be used in the design phase for sizing the links, drives and bearings of a robot.

In order to introduce required parameters for the derivation of the dynamic equations Fig. B.1 is given.

B.3.1 Recursive Newton-Euler formulation

The recursive formulation of the algorithm is mainly applied for the dynamic analysis of serial manipulators. It consists of two steps. In the forward computation phase the generalized velocities and accelerations of all links are calculated. In the backward computation step the forces and moments in all joints are determined. As in case of the Lagrangian approach the effects of the drive system, especially that of the rotor are not considered. If necessary they can be introduced as additional summands in the following equations [SSVO09].

Forward computation: First the generalized velocities of each link are computed in terms of its preceding link. They can be calculated in a recursive manner, starting at the first moving link and ending at the end-effector (TCP). As the initial condition for the base link[3]

$$^{0}\mathbf{v}_{0} = {}^{0}\dot{\mathbf{v}}_{0} = {}^{0}\boldsymbol{\omega}_{0} = {}^{0}\dot{\boldsymbol{\omega}}_{0} = \mathbf{0} \tag{B.10}$$

holds. The required relationships for the propagation of the angular velocity $^{i}\boldsymbol{\omega}_{i}$ and acceleration $^{i}\dot{\boldsymbol{\omega}}_{i}$ as well as for the linear velocity $^{i}\mathbf{v}_{i}$ and acceleration $^{i}\dot{\mathbf{v}}_{i}$, set

---

[3]In order to support the recursive formulation 0 is used instead of B for the base.

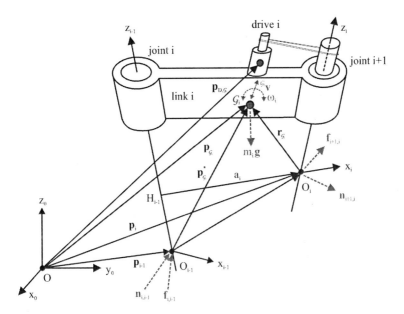

Figure B.1: Forces and moments on link i – including actuation [Tsa99, SS05]

up for each link i with respect to its preceding link $i-1$, can be found in [Tsa99]. Beside the relationship for the origin O_i of each frame i also the linear accelerations of their centers of masses $^i\dot{\mathbf{v}}_{\mathcal{G}_i}$ as well as the acceleration of gravity $^i\mathbf{g}$ have to be determined.

Backward computation: When the generalized velocities and accelerations of the links are determined the joint torques can be computed starting at the end-effector and ending at the base. As initial conditions the forces $^n\mathbf{f}_{n+1,n}$ and torques $^n\mathbf{n}_{n+1,n}$ at the end-effector have to be known. For the recursive calculation of the joint torques the following relationships hold

$$^i\mathbf{f}_{i,i-1} = {}^i\mathbf{f}_{i+1,i} - m_i \cdot {}^i\mathbf{g} + m_i \cdot {}^i\dot{\mathbf{v}}_{\mathcal{G}_i} \tag{B.11a}$$

$$^i\mathbf{n}_{i,i-1} = {}^i\mathbf{n}_{i+1,i} + \left({}^i\mathbf{r}_i + {}^i\mathbf{r}_{\mathcal{G}_i}\right) \times {}^i\mathbf{f}_{i,i-1} - {}^i\mathbf{r}_{\mathcal{G}_i} \times {}^i\mathbf{f}_{i+1,i} + {}^i\mathbf{I}_i \cdot {}^i\dot{\boldsymbol{\omega}}_i \tag{B.11b}$$
$$+ {}^i\boldsymbol{\omega}_i \times \left({}^i\mathbf{I}_i \cdot {}^i\boldsymbol{\omega}_i\right).$$

In a further step the torques are converted from the ith link frame into the $i-1$th link frame by [Tsa99]

$$^{i-1}\mathbf{f}_{i,i-1} = {}^{i-1}\mathbf{R}_i \cdot {}^{i}\mathbf{f}_{i,i-1} \tag{B.12a}$$

$$^{i-1}\mathbf{n}_{i,i-1} = {}^{i-1}\mathbf{R}_i \cdot {}^{i}\mathbf{n}_{i,i-1} \tag{B.12b}$$

Joint torques: The actuator torques $\boldsymbol{\tau}_i$ are obtained by projecting the constraint forces onto their corresponding joint axes, whereas also joint friction torques can be considered [Tsa99].

B.3.2 Application to parallel robots

In oder to apply the Newton-Euler approach to parallel robots the procedure described in [Tsa99] is used in the most cases. Primarily, it bases on a decomposition of the robot in its limbs and moving platform. The individual steps are:

1. Performing an inverse kinematic analysis in order to describe the positions, velocities and accelerations in terms of the prescribed motion of the moving platform (which means in dependency of its coordinates).

2. Decomposing the manipulator in its limbs and moving platform by cutting free at the joints connecting them. At the connection points action and reaction forces are introduced.

3. Each limb is considered as a subsystem and its balances of forces and moments are derived. Normally, in this steps some of the reaction forces can be determined independently of the equations of motion of the moving platform.

4. The remaining reaction forces are calculated by formulating the Newton and Euler equations of motion of the moving platform.

5. In the last step the actuator forces $\boldsymbol{\tau}_i$ are determined.

A detailed description of all steps can be found in [Tsa99].

B.4 Principle of virtual power

B.4.1 Theoretical background

For a system of N_k bodies the following equation can be set up applying Newton's law

$$\sum_{i=1}^{N_k} \delta \dot{\boldsymbol{\mathcal{X}}}_i^T \begin{bmatrix} m_i \ddot{\mathbf{x}}_i - m_i \mathbf{g} - \mathbf{f}_i - \mathbf{f}_{C_i} \\ \mathbf{I}_i \dot{\boldsymbol{\omega}}_i + \boldsymbol{\omega}_i \times \mathbf{I}_i \boldsymbol{\omega}_i - \mathbf{n}_i - \mathbf{n}_{C_i} \end{bmatrix} = 0, \tag{B.13}$$

where $\boldsymbol{\mathcal{X}}_i$ defines the generalized position vector of the ith body. Introducing an inertia wrench $\boldsymbol{\mathcal{F}}_i^{inertia}$, a wrench of the applied torques $\boldsymbol{\mathcal{F}}_i^{applied}$ and one for the torques of constraints $\boldsymbol{\mathcal{F}}_i^C$

$$\boldsymbol{\mathcal{F}}_i^{inertia} := \begin{bmatrix} m_i\ddot{\mathbf{x}}_i - m_i\mathbf{g} \\ \mathbf{I}_i\dot{\boldsymbol{\omega}}_i + \boldsymbol{\omega}_i \times \mathbf{I}_i\boldsymbol{\omega}_i \end{bmatrix}, \quad \boldsymbol{\mathcal{F}}_i^{applied} := \begin{bmatrix} \mathbf{f}_i \\ \mathbf{n}_i \end{bmatrix}, \quad \boldsymbol{\mathcal{F}}_i^C := \begin{bmatrix} \mathbf{f}_{C_i} \\ \mathbf{n}_{C_i} \end{bmatrix} \tag{B.14}$$

Eq. (B.13) can be simplified to

$$\sum_{i=1}^{N_k} \delta\dot{\boldsymbol{\mathcal{X}}}_i^T \left(\boldsymbol{\mathcal{F}}_i^{inertia} - \boldsymbol{\mathcal{F}}_i^{applied} - \boldsymbol{\mathcal{F}}_i^C \right) = 0. \tag{B.15}$$

In general, all torques of constraint $\boldsymbol{\mathcal{F}}_i^C$ are equal to zero, if only ideal ties (without friction) exist in a system [Hag06]. Beside for frictionless rigid body systems, this assumption does also hold for some types of friction (e.g. not in case of sliding friction torques [GPS02]). Assuming ideal ties Eq. (B.15) is reduced to

$$\sum_{i=1}^{N_k} \delta\dot{\boldsymbol{\mathcal{X}}}_i^T \left(\boldsymbol{\mathcal{F}}_i^{inertia} - \boldsymbol{\mathcal{F}}_i^{applied} \right) = 0. \tag{B.16}$$

Eq. (B.16) holds for any virtual velocity and is called *principle of virtual velocities*, *principle of virtual power* or *Jourdain's principle*. Also the special case of statics is covered in Eq. (B.16), when all acceleration are set to zero.

The equation of virtual power (B.16) can be transformed into that of virtual work by replacing the virtual velocities $\delta\dot{\boldsymbol{\mathcal{X}}}_i$ by equivalent virtual displacements $\delta\boldsymbol{\mathcal{X}}_i$. Due to the fact, that the physical dimensions as well as the magnitude of the virtual velocities are unimportant [Hag06] both equations are equivalent.

The products in Eq. (B.16) have to have the dimension of power [Moo04], whereas its individual factors not necessarily have to have the dimension of velocity/angular velocity and force/torque.

While using the virtual principles (work or power) the bodies of mechanism need not to be cut free and the forces of constraint are not occurring in the equations of motion (and have therefore not to be eliminated). In general this methods are more efficient than the Newton-Euler or Lagrangian formulation and often used in case of real-time control of parallel robots [ZS93].

B.4.2 Principle of virtual power in parameter linear form

In practice it is often advantageous to divide a parallel robot into its limbs (serial chains) which are closed over the moving platform – cp. Sec. B.3.2. Based on this basic approach the dynamic equations of parallel robots are derived in [Gro03, GHA04b, AGH05, Abd07]. The main advantage of the strategy proposed there (and recapitulated below) is the fact, that from the finally obtained equation structure the parameter vector of the parameter linear model can be reduced to its minimal form following a very systematic way. With a parameter linear and minimal model linear estimators can be applied for parameter estimation (cp. Sec. 6.1.3) and adaptive control concepts [SL88, Pie03] be realized.

The formulation of the dynamic relationships in terms of body fixed coordinate systems forms the base in order to derive the parameter-linear model [AGH05]

$$\boldsymbol{\tau} = \sum_{i=1}^{N_k} \left[\mathbf{J}_{T_i}^T \left(m_i \, {}^i\dot{\mathbf{v}}_i + {}^i\dot{\tilde{\boldsymbol{\omega}}}_i \, \mathbf{s}_i + {}^i\tilde{\boldsymbol{\omega}}_i \, {}^i\tilde{\boldsymbol{\omega}}_i \, \mathbf{s}_i \right) + \mathbf{J}_{R_i}^T \left({}^i\mathbf{I}_i \, {}^i\dot{\boldsymbol{\omega}}_i + {}^i\tilde{\boldsymbol{\omega}}_i \left({}^i\mathbf{I}_i \, {}^i\boldsymbol{\omega}_i \right) + {}^i\tilde{\mathbf{s}}_i \, {}^i\dot{\mathbf{v}}_i \right) \right],$$
$$(B.17)$$

where $\mathbf{s}$ is the vector of the first moments, whose elements are defined by $\mathbf{s}_i = m_i \, {}^{i,\mathcal{G}_i}\mathbf{r}_i$ ($^{i,\mathcal{G}_i}\mathbf{r}_i$: vector from coordinate frame i to $\mathcal{G}_i$) and the tilde-operator $\tilde{(\cdot)}$ defines the cross-product of the two vectors $\tilde{a}b = a \times b$. The definitions of the translation and rotational Jacobians can be found in [AGH05]. With the operators $(\cdot)^\circledast$ and $(\cdot)^\circ$ defined in [CB97, GJKHG02], which fulfills

$$\boldsymbol{\omega}_i^\circledast \, \mathbf{I}_i^\circ := {}^i \mathbf{I}_i \, {}^i\boldsymbol{\omega}_i \qquad (B.18)$$

with

$$\boldsymbol{\omega}_i^\circledast := \begin{bmatrix} \omega_{i,x} & \omega_{i,y} & \omega_{i,z} & 0 & 0 & 0 \\ 0 & \omega_{i,x} & 0 & \omega_{i,y} & \omega_{i,z} & 0 \\ 0 & 0 & \omega_{i,x} & 0 & \omega_{i,y} & \omega_{i,z} \end{bmatrix} \qquad \text{and}$$

$$\mathbf{I}_i^\circ = \begin{bmatrix} I_{i,xx} & I_{i,xy} & I_{i,xz} & I_{i,yy} & I_{i,yz} & I_{i,zz} \end{bmatrix}^T$$

the rigid body dynamic equations can be simplified to [AGH05]

$$\boldsymbol{\tau}_{rb} = \sum_{i=1}^{N_k} \underbrace{\begin{bmatrix} \mathbf{J}_{T_i}^T & \mathbf{J}_{R_i}^T \end{bmatrix} \boldsymbol{\Omega}_i}_{\mathbf{H}_i} \underbrace{\begin{bmatrix} \mathbf{I}_i^\circ \\ \mathbf{s}_i \\ m_i \end{bmatrix}}_{\mathbf{p}_i} = \underbrace{\begin{bmatrix} \mathbf{H}_1 & \cdots & \mathbf{H}_{N_k} \end{bmatrix}}_{\mathbf{H}_{rb}(\mathbf{q},\dot{\mathbf{q}},\ddot{\mathbf{q}})} \underbrace{\begin{bmatrix} \mathbf{p}_1 & \cdots & \mathbf{p}_{N_k} \end{bmatrix}^T}_{\mathbf{p}_{rb}} \qquad (B.19)$$

with

$$\boldsymbol{\Omega}_i = \begin{bmatrix} 0 & {}^i\dot{\tilde{\boldsymbol{\omega}}}_i + {}^i\tilde{\boldsymbol{\omega}}_i \, {}^i\tilde{\boldsymbol{\omega}}_i & {}^i\dot{\mathbf{v}}_i \\ {}^i\dot{\boldsymbol{\omega}}_i^\circledast + {}^i\tilde{\boldsymbol{\omega}}_i \, {}^i\tilde{\boldsymbol{\omega}}_i^\circledast & -{}^i\dot{\mathbf{v}}_i & 0 \end{bmatrix} .$$

In terms of the actuator torques, the rigid body dynamics can be expressed in the following parameter-linear form

$$\mathcal{F}^a_{rb} = \mathbf{J}^T \boldsymbol{\tau}_{rb} = \mathbf{J}^T \mathbf{H}_{rb} \, \mathbf{p}_{rb}, \tag{B.20}$$

where $\mathbf{J}$ is the Jacobian matrix of the parallel robot. The models in Eq. (B.19) and (B.20) are not necessarily parameter-minimal. But they can be reduced to this form in a very systematic way [AGH05].

B.5 Torques and friction torques in detail

Several groups of torques are acting on a robot: the gravitational forces $\mathcal{F}^g$, the inertia torques $\mathcal{F}^{inertia}$, the joint torques produced by the actuators $\mathcal{F}^a$, friction torques $\mathcal{F}^{friction}$ and external torques $\mathcal{F}^{ext}$. Depending on the applied modeling strategy some of these torques get combined, whereas the others are considered separately. Especially $\mathcal{F}^g$ is in the most cases, and also here, considered as part of $\mathcal{F}^{inertia}$, cp. Eq. (B.14), and not listed separately. Furthermore, the applied torques $\mathcal{F}^{applied}$ are in general defined as the sum of the actuator $\mathcal{F}^a$ and external torques $\mathcal{F}^{ext}$

$$\mathcal{F}^{applied} = \mathcal{F}^a + \mathcal{F}^{ext} \tag{B.21}$$

whereas it is commonly assumed, that the external torques are only exerted at the end-effector.

The generalized torque $\mathcal{F}$ is introduced in order to summarize all torques acting on a robot which are consistent with its mechanical constraints. Applying the principle of virtual work the vector can be expressed by

$$\delta\mathcal{W} = \mathcal{F}^T \, \delta\mathbf{q} \tag{B.22}$$

In cases where actuators exert torques at the joints and external torques are exerted at the end-effector, the produced virtual work is given by [Tsa99]

$$\delta\mathcal{W} = \left(\mathcal{F}^a\right)^T \delta\mathcal{X} + \left(\mathcal{F}^{ext}\right)^T \delta\mathcal{X} \tag{B.23}$$

where $\mathcal{F}^a$ is the $(n \times 1)$-vector of the joint torques generated by the actuators, $\mathcal{F}^{ext}$ a (6×1)-vector of the resultant torques at the end-effector and $\delta\mathcal{X}$ the (6×1)-vector of the virtual displacement of the end-effector. These are related to the virtual displacements of the active joints [Tsa99] by

$$\delta\mathcal{X} = \mathbf{J}^{-1} \, \delta\mathbf{q}, \tag{B.24}$$

where $\mathbf{J}$ is the Jacobian matrix. Substituting Eq. (B.24) into Eq. (B.23) and equating Eq. (B.23) and Eq. (B.22) yields[4]

$$\mathcal{F} = \mathbf{J}^{-T} \, \mathcal{F}^{a}_{(rb)} + \mathbf{J}^{-T} \, \mathcal{F}^{ext} = \mathbf{J}^{-T} \, \mathcal{F}^{applied}, \tag{B.25}$$

where the vector of the actuator torques is defined by

$$\boldsymbol{\tau}_{(rb)} = \mathbf{J}^{-T} \, \mathcal{F}^{a}_{(rb)}. \tag{B.26}$$

The contribution of friction to $\mathcal{F}$ can also be formulated. The friction torques $\mathcal{F}^{friction}$ are normally highly non-linear, difficult to model accurate and have a significant effect on system dynamics [Tsa99]. Their contribution is given by

$$\delta \mathcal{W} = - \left(\mathcal{F}^{friction} \right)^{T} \delta \mathcal{X}, \tag{B.27}$$

where $\mathcal{F}^{friction}$ is the vector of the friction torques and the minus sign indicates, that its direction is always opposite to that of the joint movements. The extension of Eq. (B.25) by a friction term yields

$$\mathcal{F} = \mathbf{J}^{-T} \left(\mathcal{F}^{a}_{(rb)} + \mathcal{F}^{ext} - \mathcal{F}^{friction} \right). \tag{B.28}$$

In the most cases viscous $\mathbf{F}_{v}\dot{\mathbf{q}}$ and static friction torques $\mathbf{F}_{s}\mathbf{sgn}(\dot{\mathbf{q}})$ (here: Coulomb friction) are considered

$$\mathcal{F} = \boldsymbol{\tau}_{(rb)} + \mathbf{J}^{-T} \, \mathcal{F}^{ext} - \mathbf{F}_{v}\dot{\mathbf{q}} - \mathbf{F}_{s} \, \mathbf{sgn}(\dot{\mathbf{q}}), \tag{B.29}$$

where $\mathbf{F}_{v}$ denotes the $(n \times n)$ diagonal matrix of viscous friction coefficients, $\mathbf{F}_{s}$ a $(n \times n)$ diagonal matrix, $\mathbf{sgn}(\dot{\mathbf{q}})$ a $(n \times 1)$ vector whose components are given by the sign functions of the generalized joint velocities [SSVO09]. In absence of friction and externally applied torques (no contact to the environment) the vector of generalized forces is equivalent to that of the generalized joint forces [Tsa99]

$$\mathcal{F} = \boldsymbol{\tau}_{(rb)}. \tag{B.30}$$

Which means that the components of $\mathcal{F}$ are the actuator forces of the prismatic joints and the torques of the revolute ones. In case of the here performed dynamic modeling assumption (B.30) is valid during the modeling and comparison of system dynamics in Chap. 5. In case of the parameter-identification in Chap. 6 the models are extended by active and passive joint friction.

---

[4]The notation (rb) is introduced to separate the rigid body and friction parts of the torques.

C Basic strategies for trajectory planning and trajectories

In this section commonly known strategies for trajectory planning are shortly summarized. The purpose of this chapter is to recapitulate some facts needed in Chap. 7 for introducing the new strategy and to present all of the determined trajectories of the SpiderMill in a compact form.

C.1 Classical strategies for trajectory planning

C.1.1 Basic algorithm

Almost simultaneously a basic algorithm for time-optimal trajectory planning has been developed by the work groups of Bobrow [BDG85], Shin [SM85] and Pfeiffer and Johanni [PJ86]. The basic idea of all three algorithms is the description of the robot dynamics as a function of the path parameter. Exemplarily the version of Pfeiffer and Johanni is discussed in more detail.

Concept: The algorithm [PJ86, PJ87] can be applied to any kind of manipulator but takes only the path and torque CTs into account, strongly simplifying the optimization problem (7.20). It threats the MVC – which is equivalent to the T-velocity-limit curve of Sec. 7.1.3 as source and sink for the trajectories. Below this curve a trajectory field is calculated based on the torque limits. The approach is very descriptive working in the space of the path variable s in which the torque limits are transformed. For a fixed s an AR like that in Fig. 7.2 (left) is obtained in the $\dot{s}$-$\ddot{s}$ phase plane. The region limits $\ddot{s}$ for a fixed s and $\dot{s}$, cp. Eq. (7.27). Furthermore from the AR an upper bound for $\dot{s}$ can be identified, cp. Eq. (7.29).

MVC: Each point on the MVC in the s-$\dot{s}$ phase plane – except the critical ones – is characterized by exactly one value of $\ddot{s}$, cp. Fig. 7.2 (left). Each value of $\dot{s}$ below the MVC features two extrema of the $\ddot{s}$. These extrema are the minimal or maximal reachable $\ddot{s}$ in the point $(s, \dot{s})$. By means of both extrema a gradient field can be constructed below the MVC. This characterizes the course of possible trajectories. In Fig. C.1 a MVC, a part of the gradient field as well as the belonging trajectories are exemplarily shown.

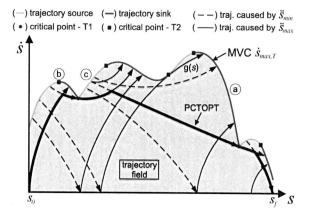

Figure C.1: Approach of Pfeiffer and Johanni in the s-$\dot{s}$ phase plane

Sink and source: The trajectories can be constructed based on

states: $x_1 = s$, $x_2 = \dot{s}$ input: $u = \ddot{s} \in \{\ddot{s}_{min,T}(x_1, x_2), \ddot{s}_{max,T}(x_1, x_2)\}$

$$\text{model:} \begin{bmatrix} \dot{x}_1 \\ \dot{x}_2 \end{bmatrix} = \begin{bmatrix} x_2 \\ 0 \end{bmatrix} + \begin{bmatrix} 0 \\ 1 \end{bmatrix} \cdot u, \tag{C.1}$$

where $\ddot{s}_{min,T}$ and $\ddot{s}_{max,T}$ are calculated with Eq. (7.27). Along the plotted trajectories $\ddot{s}$ is maximal or minimal. Since the limits of $\ddot{s}$ are a result of the torque limits of the drives, at each time instant at least one drive is on its torque limit along the shown trajectories. Therefore the trajectories of the trajectory field in Fig. C.1 feature a time minimizing characteristic.

The gradient field and the belonging trajectories are used to construct a path-constrained time optimal trajectory (PCTOPT). Therefor a further analysis of the gradient field and the MVC is necessary. From Fig. C.1 it can be seen, how a trajectory source and sink acts. This means, that trajectories can be constructed from the gradient field, which either starts or ends on the MVC. The reason for it are the values of $\ddot{s}$ on the MVC. Sole exception are the points on the MVC, in which its characteristics changes form a trajectory source into a sink or the other way around, cp. Fig. C.1. In this points a trajectory, which is constructed from the gradient field, can be a tangent to the MVC. These points are called critical points and play an important role in the design process of PCTOPTs.

Critical point: A critical point is a transition between a trajectory source and

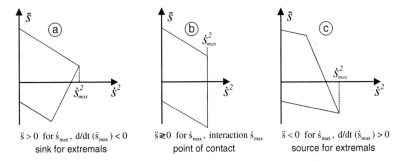

$\ddot{s} > 0$ for $\dot{s}_{max}$, $d/dt\,(\dot{s}_{max}) < 0$
sink for extremals

$\ddot{s} \gtreqless 0$ for $\dot{s}_{max}$, interaction $\dot{s}_{max}$
point of contact

$\ddot{s} < 0$ for $\dot{s}_{max}$, $d/dt\,(\dot{s}_{max}) > 0$
source for extremals

Figure C.2: Trajectory sources, sinks and contact (critical) points [PJ87]

sink, where the maximal pseudo-velocity $\dot{s}_{max}^2$ is not uniquely defined. It arises, when in the $\dot{s}$-$\ddot{s}$ phase plane in a point $\dot{s}_{max}^2$ the pseudo-acceleration $\ddot{s}$ is indeterminated, cp. Fig.C.2 case ⓑ. For that case the intersection line is a parallel to the ordinate, whereas in case of a sink ⓐ or a source ⓒ only one value is valid. Two types of critical points can be distinguished. In case of the first type (T1)

$$m_i(s) = 0 \qquad (C.2)$$

holds. The MVC is at this point non-differentiable with respect to s. In case of the second type (T2) the trajectory $g(s)$ constructed from the gradient field becomes a tangent to the MVC $\dot{s}_{max}(s)$ in a point in which it is differentiable. In this point both have the same slope, consequently the sign of

$$\kappa(s) = \frac{d\dot{s}_{max}(s)}{ds} - \frac{dg(s)}{ds} \qquad (C.3)$$

changes [SM83, SM85]. In this point the trajectory meets the MVC without violating the actuator constraints. Based on the knowledge of the critical points a time-optimal trajectory can be constructed.

Design of the trajectory: The strategy is explained based on Fig. C.1. Starting point are the MVC and the knowledge of the critical points. Form the start point s_0 it will be forward integrated with maximal pseudo-acceleration and from the end point s_f backward with minimal pseudo-acceleration. Thereto the model in Eq. (C.1) is used. If both trajectories intersect below the MVC the search for the time-optimal trajectory is finished. In the other case, when both trajectories intersects the MVC, further trajectories in between have to be found. Therefor the critical points between the two points of intersection are used. They define the transition between trajectory source and sink and a trajectory will only be a tangent to it. From each of the critical points it will be forward integrated with maximal pseudo-acceleration and backward with minimal one. By this procedure a way from the starting s_0 to the end point s_f is obtained, which is equivalent to the time-optimal trajectory. Beyond the base algorithm, also further considerations like smoothing the trajectory by introducing further optimization criteria based on a cost function are discussed in [PJ87].

One point has bot be emphasized: A time-optimal trajectory can not completely take course along the MVC, since the MVC does not considers all dynamic boundary conditions [Pie03]. In general a time-optimal trajectory can only follow the MVC in single points or intervals. For that reason it is important to find the reachable points on the MVC – different strategies for solving this problem have been developed by various research groups. The most detailed differentiation of the different kinds of critical points can be found in [SY89].

C.1.2 Extensions of the basic algorithm

Algorithm without MVC considering friction: An extension additionally considering friction terms and avoiding the calculation of the MVC has been proposed by Slotine and Yang [SY89]. Therefor the definition of the critical points has been extended. The authors distinguish three types of characteristic switching points: zero-inertia points ($m_i(s) = 0$), tangent points and discontinuity points (effect of the friction terms). Furthermore a strategy for determining them directly has been introduced.

Application to parallel robots: Abdellatif and Heimann [AH05] has applied the base strategy of Slotine and Yang for the first time to a parallel robot additionally considering further friction influences. In order to handle the resulting large number of critical points in case of parallel robots – caused by the several limbs (each limb introduces own candidates for switching points) and the friction – and therefore large number of required integrations an algorithm for sorting the critical point by their $\dot{s}$ has been introduced. A selection of critical points out of the ordered set is used in a further step to construct the time-optimal trajectory. Also in this approach only actuator bounds are considered, neglecting a jerk limitation.

Planning of smooth trajectories: The basic algorithm only considers path and torque CTs. The constructed trajectories are indeed time-optimal but not smooth. Therefore the approach has been enhanced by Constantinescu and Croft [Con98, CC00] for the planning of smooth PCTOPT ($\mapsto$ PCSMTOPT). The strategy starts with the time-optimal trajectory obtained by the basic approach and bases on so called transitions. These transitions define intervals in which cubic splines are used for the approximation of the PCSMTOPT. The height of the points (in the phase-plane) as well as the slope in the starting- and end-point of the trajectory become parameters of the optimization problem. Beside limitations for the path and the torques, the approach also considers constraints for drive jerks. Limits for the velocities of the active joints are not defined. This leads to a simplification of the optimization problem in Eq. (7.20), due to the reduced set of CTs. The physical limits of the torque and jerk are transformed into the state-space of s. As a result the ARs in the $\dot{s}$-$\ddot{s}$- and $\ddot{s}$-$\dddot{s}$-phase plane (cp. Fig. 7.2) are obtained. From the planes limits for $\dot{s}$, $\ddot{s}$ and $\dddot{s}$ can be derived (cp. Sec. 7.1.3), where in contrast to Sec. 7.1.3 the limits for $\dot{s}$ are not influenced by a velocity CT.

The design of the PCSMTOPTs is performed in the s-$\dot{s}$ phase plane, since here s_0 and s_f are fixed. If the time would be used as variable instead, the end-point t_f would be free and influenced by the optimization.

The term 'smooth' says, that the $s(t)$, $\dot{s}(t)$ and $\ddot{s}(t)$ are continuous functions. This is guaranteed by the limitation of $\dddot{s}(t)$. Thus [PJ87]

$$\ddot{s} = \frac{d\dot{s}}{ds} \cdot \dot{s} \quad \Leftrightarrow \quad \frac{d\dot{s}}{ds} = \frac{\ddot{s}}{\dot{s}} \qquad \text{(C.4)}$$

is valid. Since $\dot{s}(t)$ and $\ddot{s}(t)$ are continuous functions, also $\dot{s}(s)$ in the s-$\dot{s}$ phase plane is a continuous and differentiable function. By the design of a smooth function in the s-$\dot{s}$ phase plane hence a smooth trajectory is obtained.

But the trajectory has also to be time-optimal, which implies, that the time t_f for which the end-point $s(t_f)$ is reached has to be minimized. For representing the requirement of time-optimality in the s-$\dot{s}$ phase plane the following relationship [PJ87]

$$\dot{s}(s) = \frac{ds}{dt} \quad \Rightarrow \quad t_f = \int\limits_{s_0}^{s_f} \frac{1}{\dot{s}(s)} ds. \qquad \text{(C.5)}$$

is used. From it it can bee seen, that the time t_f is minimized, when in the s-$\dot{s}$ phase plane the area below the function $\dot{s}(s)$ is maximized.

For the design of the trajectories in the s-$\dot{s}$ phase plane trial functions are used. The abscissa therefore is divided into intervals in which the trial functions lie. For an exact solution of the optimization problem (7.20) (without velocity CT) an infinite number of intervals would be necessary. But for an approximation of the optimal solution a finite number of intervals is sufficient. In [CC00] cubic splines are suggested as trial functions. They are smooth, differentiable and simple to parametrize. Since the curves which are generated by the splines shall be smooth in the whole s-$\dot{s}$ phase plane and thus also at the transitions of the intervals the values at the interval transitions $\dot{s}_i$ as well as the slopes in the starting- $\frac{d\dot{s}}{ds}$ (s_0) and end-point $\frac{d\dot{s}}{ds}$ (s_f) get free selectable parameters and consequently the optimization variables.

What remains is the choice of the intervals. For this the strategy of Pfeiffer and Johanni [PJ87] is used. Therein $|\dot{\tau}| \to \infty$ is used as 'limit' for the jerk. This means, the trajectory of the basic approach gets a limit curve for the new one. Consequently it is suggestive to use as transitions between the intervals the critical points, as well as the not differentiable transitions along the trajectory, cp. Fig. C.3. In [CC00] it is also shown, that additional transitions only lead to a

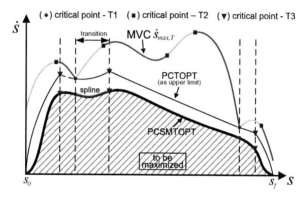

Figure C.3: Transitions and smooth, time-optimal trajectory

considerable reduction of the end time t_f if their number has been to low so far. The optimization problem is solved using numerical optimizers. The resulting trajectory is time-optimal and smooth under the introduced approximations.

C.2 Trajectories planned for the SpiderMill

In Fig. C.4 to C.10 the trajectories planned for the SpiderMill with the new strategy proposed in Chap. 7 are given. Based on the proposed strategy they have been calculated in [M08].

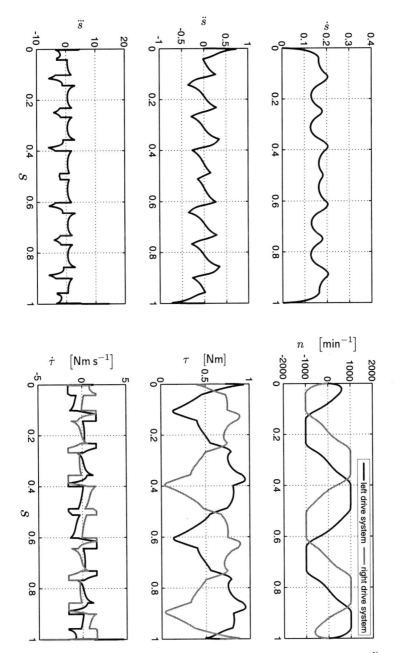

Figure C.4: PCSMTOPT for the circle path with a maximal jerk of $|\dot{\tau}| \leq 1.6 \, \frac{\text{Nm}}{\text{s}}$

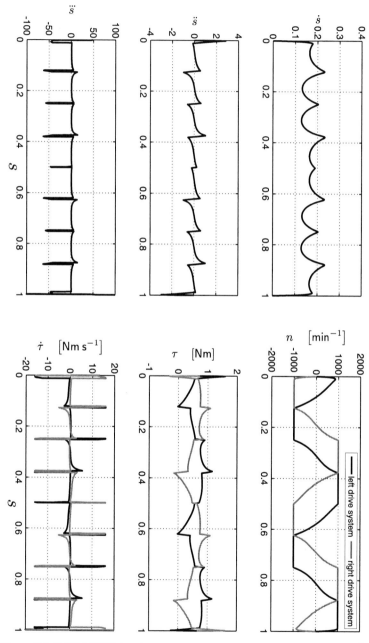

Figure C.5: PCSMTOPT for the circle path with a maximal jerk of $|\dot{\tau}| \leq 16 \, \frac{\mathrm{Nm}}{\mathrm{s}}$

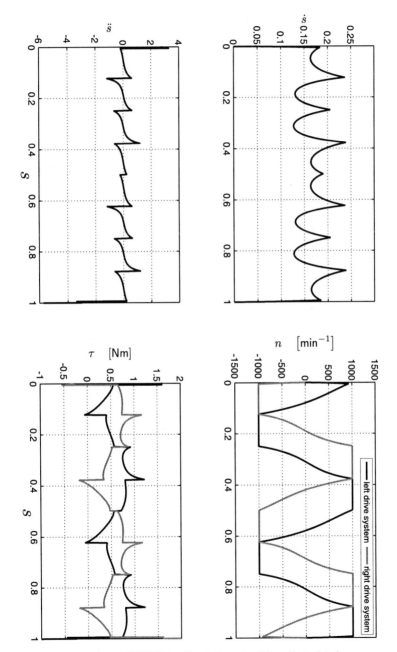

Figure C.6: PCTOPT for the circle path with unlimited jerk

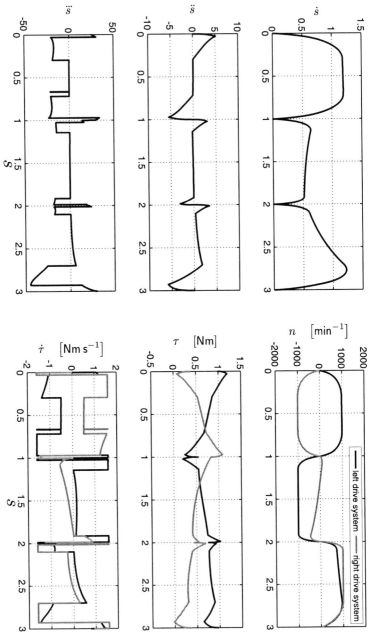

Figure C.7: PCSMTOPT for the triangle path with a maximal jerk of $|\dot{\tau}| \leq 1.6\,\frac{\mathrm{Nm}}{\mathrm{s}}$

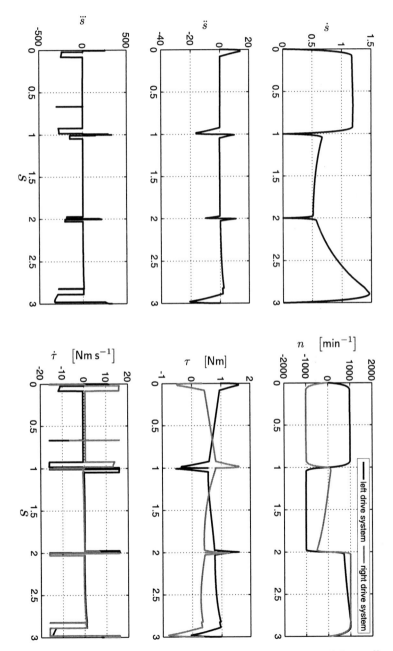

Figure C.8: PCSMTOPT for the triangle path with a maximal jerk of $|\dot{\tau}| \leq 16 \, \frac{\mathrm{Nm}}{\mathrm{s}}$

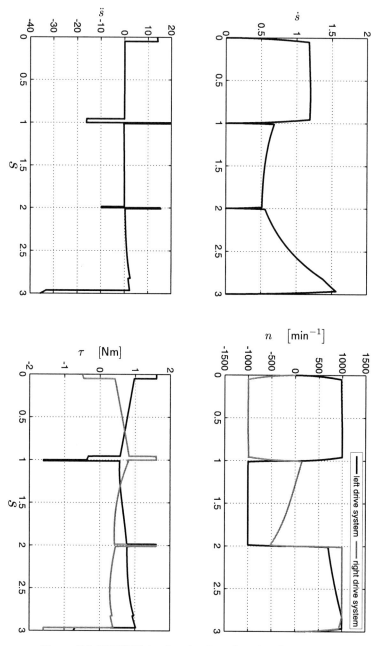

Figure C.9: PCTOPT for the triangle path with unlimited jerk

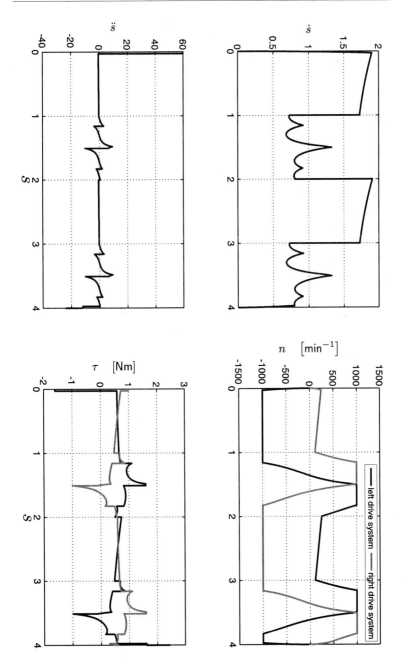

Figure C.10: PCTOPT for the eight path with unlimited jerk

D Nomenclature

mathematical abbreviations and definitions

$(\dot{\cdot}) = \partial/\partial t$ $\qquad$ $(\cdot)' = \partial/\partial s$ $\qquad$ $(\hat{\cdot})$: uncertain/observed value

$(\tilde{\cdot}) = (\cdot) - (\hat{\cdot})$ $\qquad$ $\overline{(\cdot)}$: upper limit $\qquad$ $\underline{(\cdot)}$: lower limit

$()_{min}$: minimal value $\quad$ $()_{max}$: maximal value $\quad$ $L_f h(\mathbf{x}) = \dfrac{\partial \mathbf{h}}{\partial \mathbf{x}} \mathbf{f}(\mathbf{x})$

$s(\alpha) = \sin(\alpha)$ $\qquad$ $c(\alpha) = \cos(\alpha)$ $\qquad$ $t(\alpha) = \tan(\alpha)$

coordinates, coordinate systems and planes

$(\cdot)_{TCP,A}$ $\qquad$ TCP in MSC.ADAMS coordinate system

$^B(\cdot)$ $\qquad$ base coordinate system (simplified structure)

$(\cdot)_{TCP,B}$ $\qquad$ TCP for simplified structure

$^{BI}(\cdot)$ $\qquad$ base coordinate system of the inner loop

$(\cdot)_{TCP,BI}$ $\qquad$ TCP in case of the inner loop

$^i(\cdot)$ $\qquad$ coordinate system i as reference frame

x_i, y_i $\qquad$ Cartesian coordinates

E_V $\qquad$ virtual plane

indices

b	back loop	f	front loop	m	middle (between b and f)
k	side (general)	l	left side	r	right side
B	base	CM	moving platform	E	end-effector
x	state-space of $\mathbf{x}$	z	state-space of $\mathbf{z}$	d	desired/reference value
J	limitation as consequence of jerk limits				
T	limitation as consequence of torque limits				
V	limitation as consequence of velocity limits				

joint variables and vectors of joint variables

γ_k passive joint variable (angle)

s_k active joint variable: spindle length

δs change of spindle-length

θ_k pseudo-active joint-variable (angle)

q_i generalized joint variable

$\mathbf{q}_a$ vector of active (prismatic) joint variables: $\mathbf{q}_a = [s_l, s_r]^T$

$\mathbf{q}$ vector of generalized joint variables

$\mathbf{q}_p$ vector of passive joint variables $\mathbf{q}_p = [\gamma_l, \gamma_r]^T$

$\mathbf{q}_{pa}$ vector of pseudo-active (revolute) joint variables: $\mathbf{q}_{pa} = [\theta_l, \theta_r]^T$

geometric constants and parameters of the SpiderMill

length:	a, b, δ, d
distance to center of mass:	d_{mi}
angle:	$\alpha, \beta, \epsilon, \theta_{off}, \mu$

parameters of the drive system of the SpiderMill

A_k contact point of the spindle forces f_{sk}

D_k center of rotation of motor suspension of side k, drive of side k

h spindle slope

τ_i^{max} maximum torque of drive i

number of ...

n	point masses	N_E	the equations of motion
n_{dr}	drives	N_k, N_{k_j}	bodies of a mechanism / of its limb j
n_T	trajectories	N_p	pulses during the interval t_S
N	sampling points	N_s	serial chains of a mechanism

energy terms, kinematic functions, sets

$\mathcal{C}_A$	set of constraints of kinetic quantities	$\mathcal{L}$	Lagrangian (function)
$\mathcal{D}$	direct kinematics	$\mathcal{T}$	kinetic energy
$\mathcal{I}$	inverse kinematics	$\mathcal{U}$	potential energy

structural elements and parameters

C_i coupling element (i: general, B: base, M: moving platform)
$G_{i(k)}$ group i of links
$J_{i(k)}$ substitution joint (i: massless, mi: with point mass)
$L_{i(k)}$ link i
$L_{Gij(k)}$ link j of group i of links
$S_{i(k)}$ substitution part i
λ operational space of a mechanism
f_i degree of freedom of joint i
f_p passive/identical DOFs
F DOF of mechanism
z redundant constraints

'time' parameters

t	time variable	t_f	traveling time	t_{set}	settling time
t_i	switching instants	t_k	sampling instant	T_S	sampling time
t_{ex}	code execution time	t_r	rise time		

motion parameters

$x(t)$	position	s	path variable	$\mathbf{r}(s)$	path
$v(t)$	velocity	$\dot{s}$	pseudo-velocity	s_T	transition point
$a(t)$	acceleration	$\ddot{s}$	pseudo-acceleration	s_0/s_f	start/end point of path
$j(t)$	jerk	$\dddot{s}$	pseudo-jerk	$n(t)$	rotational speed

vectorial motion parameters

$\mathbf{a}$ acceleration of the mass m or a system of masses
$\mathbf{g}$ vector of the gravitational acceleration
$\mathbf{r}_i$ position vector i
$\mathbf{v}_{\mathcal{G}_i}$ absolute linear velocity of the center of mass of link i
$\mathbf{x}$ position of the center of gravity of the total mass m
$\mathbf{x}_i$ position vector of the point mass i
$\dot{\mathbf{v}}_{\mathcal{G}_i}$ acceleration of the center of mass of link i
$^{B}\dot{\mathbf{v}}_{mjk}$ acceleration of point mass m_j
$\boldsymbol{\omega}_{\mathcal{G}_i}$ absolute angular velocity of the center of mass of link i
$^{B}\boldsymbol{\omega}_{jk}$ angular velocity of body j
$^{B}\dot{\boldsymbol{\omega}}_{jk}$ angular acceleration of body j

scalars constants and variables

$d_{i(k)}$ — distance between two consecutive substitution joints; length of $S_{i(k)}$

$d_{mi(k)}$ — distance between joint i and the location of $m_{i(k)}$ of $S_{i(k)}$

g — gravitational acceleration constant

I — moment of inertia

$I_{mi(k),3}$ — principal moment of inertia with respect to the z-axis of the body fixed coordinate system in $m_{i(k)}$

l_i — distance between the position $\mathbf{x}_i$ of m_i and the position $\mathbf{x}$ of center of gravity of m with respect to a coordinate system in $\mathcal{G}$

m_i — point mass i, mass of link i

m^* — end-mass at a single link (point mass)

p_i — dynamic parameter of rigid body model

r_i — ith radius of rotation, relative degree of subsystem i

r_{ai} — friction coefficient actuation part i

r_{pi} — friction coefficient passive joint i

vectors of dynamic modeling and parameter identification

$\mathbf{f}_{\mathcal{G}_i}$ — force acting on $\mathcal{G}_i$ of link i

$\mathbf{f}_{jk}$ — force from body j to body $j+1$

$\mathbf{f}_{sk}$ — spindle force k

$\mathbf{G}$ — vector of gravity terms

$\mathcal{G}_i$ — center of mass of body i

$\mathbf{n}_i$ — vector of (friction) torque

$\mathbf{p}_f$ — vector of friction coefficients of the passive joints and actuation parts

$\mathbf{p}_{D,\mathcal{G}_i}$ — vector from the base frame to the center of mass of the rotor

$\mathbf{p}_{rb}$ — vector of the dynamic parameters of the rigid body model

$\mathbf{z}_{i-1}$ — unit vector pointing along the ith joint axis

$\boldsymbol{\tau}_i$ — vector of the generalized (friction) forces

matrices of dynamic modeling and parameter identification

$\mathbf{C}$	centrifugal and Coriolis terms	$\mathbf{I}_i$	inertia matrix
$\mathbf{D}_i$	transformation matrix	$\mathbf{L}$	estimator gain matrix
$\mathbf{E}$	identity matrix, see also $\mathbf{I}$	$\mathbf{M}$	mass/inertia matrix
$\mathbf{F}$	friction matrix	$\mathbf{J}$	manipulator Jacobian matrix
$\mathbf{H}$	estimator gain matrix	$\mathbf{J}_i$	link Jacobian (sub-)matrix
$\mathbf{H}_i$	regressor matrix	$\boldsymbol{\Gamma}$	measurement vector
$\mathbf{I}$	identity matrix, see also $\mathbf{E}$	$\boldsymbol{\Psi}$	information matrix

description of the dynamic systems

$\mathbf{A}$	system matrix	$\mathbf{u}$	input/control vector	d	damping factor/ratio
$\mathbf{B}$	input matrix	$\mathbf{x}$	state vector	λ_i	ith eigenvalue
$\mathbf{C}$	output matrix	$\mathbf{y}$	output vector	ω_0	natural frequency
$\mathbf{D}$	direct matrix	$\boldsymbol{\eta}$	uncertainty vector		

observer and control concepts

$\mathrm{COV}(x)$	covariance of x	$\mathbf{Q}$	weighting/covariance matrix
$e, \mathbf{e}$	error and error vector	r	reference input
$E(\,\cdot\,)$	expectation value	$\mathbf{R}$	weighting/covariance matrix
$\mathbf{F}$	feedforward compensation	T	transfer function closed loop
G	transfer function	$\mathbf{v}_k$	noise process
G_C	transfer function controller	V	prefilter
h_i	observer gain	$\mathbf{w}_k$	noise processes
$\mathbf{H}$	estimator gain matrix	$\hat{\mathbf{x}}_k^+, \hat{\mathbf{x}}_k^-$	a posteriori/priori estimate
$\mathbf{J}$	Jacobi matrix	δ	Kronecker delta function
k_i	controller gain	$\boldsymbol{\delta}$	(model) disturbance
$\mathbf{K}_f$	gain matrix LKF	φ_r	phase margin
$\mathbf{K}_r$	gain matrix LQR	$\boldsymbol{\varphi}_i$	non-linear model part
$\mathbf{L}$	estimator gain matrix	$\sigma^2(x)$	variance of x
L	transfer function open loop	$\theta_m^{(\star)}, \theta_m^{\star}$	interpulse angle/length change
$\mathbf{P}_k$	covariance of estimation error	ω_c	crossover frequency

functions, models and transformations

J	cost function	$\boldsymbol{\alpha}(\mathbf{x})$	feedback function
n_{ai}	friction model actuation part i	$\boldsymbol{\beta}(\mathbf{x})$	feedback function
n_{pi}	friction model passive joint i	Γ_i	ith constraint function
$\mathbf{T}$	transformation matrix	$\boldsymbol{\phi}(\mathbf{x})$	state transformation

performance indices

$\overline{\Delta(\cdot)}$	mean value of a performance index
η_i	drive degree of utilization for the ith trajectory
ΔP_i	average trajectory error of trajectory i
ΔS_i	average error of the spindle coordinates for the ith trajectory
$\Delta \dot{S}_i$	average error of the spindle velocities for the ith trajectory

abbreviations

AW	Adaptive Windowing
AR	admissible region
AWR	Anti-Windup-Reset
BP	branching point
CAD	computer aided/assisted design
CAS	computer algebra system
CT	constraint
CTS	constraint singularity
DOF	degree of freedom
DAE	differential algebraic equation
DDP	direct dynamic problem
FDM	Finite Difference Method
FEA	finite element analysis
FM	full model (no simplification in advance)
FOAW	First-Order Adaptive Windowing
FFM	Fixed Filter Method
I/O	input/output
I/S	input/state
IDP	inverse dynamic problem
ITA	Inverse Time Approach
lin.	linear
LKF	discrete Linear Kalman Filter
LS	least square(s)
LQE	Linear Quadratic Estimator
LQG	Linear Quadratic Gaussian
LQR	Linear Quadratic Regulator
MVC	maximal-velocity-limit-curve
MAE	mean absolute error
MRE	mean relative error
MBS	multi-body system
MIMO	multiple-input-multiple-output
NC	numerical control
n.l.	nonlinear
OAR	overall admissible region
PCSMTOPT	path-constrained, smooth, time-optimal trajectory
PCTOPT	path-constrained, time optimal trajectory
ptp	point-to-point
RTW	Real-Time-Workshop (of Matlab/Simulink)
SQP	sequential quadratic programming
SM	simplified model
SMM	SimMechanics model
SISO	single-input-single-output
SMTOPT	smooth, time-optimal trajectory
TCP	tool center point
WP	working point

E Summary

In industrial practice most robots are of serial kinematic structure. But in case of time-sensitive, highly dynamical applications or processes requiring a high structural rigidity of the robots they are more and more replaced by parallel kinematic manipulators.

In this thesis a holistic concept for the model-based trajectory tracking control of a special class of these robots has been introduced and demonstrated at the planar parallel robot SpiderMill. The studied class is that of parallel kinematic structures containing functional or structural redundancies (complex parallel robots). Since for these robots – due to the high complexity of their dynamic models – sophisticated model-based strategies for trajectory tracking control cannot be directly applied, a new concept was necessary and has been introduced here. It mainly bases upon the derivation of real-time implementable compact dynamic models for the studied class of robots. But compact dynamic models are only a first step on the way to a model-based trajectory tracking control with a high path accuracy and a low traveling time. Additionally, appropriate planned trajectories are required. For planning them a new strategy has been introduced.

In the following all performed steps on the way to the model-based trajectory tracking control are shortly summarized: After introducing the planar parallel robot SpiderMill (Chap. 2) and performing a structural analysis including its DOFs, workspace, singularities as well as its direct and inverse kinematics (Chap. 3), a new strategy for modeling of parallel robots with redundancies has been introduced and applied to the SpiderMill (Chap. 4). Based on the simplified structure of the robot several standard strategies for dynamic modeling have been used in order to derive analytical and real-time implementable dynamic models (Chap. 5). For one of these models all further steps in terms of model-based trajectory tracking control have been performed, starting with the identification of its rigid body and friction parameters (Chap. 6), to further improve the model quality. Since in literature, only a few trajectory planning strategies – with several disadvantages in case of complex parallel robots – can be found, a new approach for the planning of path-constrained smooth trajectories (PCSMTOPTs) has been introduced in the next step (Chap. 7).

Based on the feedback linearization of the non-linear system model (Chap. 8) it has been possible to apply linear control theory for the design of the trajectory tracking controllers. Since for the feedback-linearization and the most control concepts information about all system states are required – which means especially that of the joint or spindle velocities – possible strategies for their determination (observation) have been analyzed in a following step (Chap. 9). Finally model- and non-model-based control concepts have been implemented and tested in a simulation environment as well as at the physical demonstrator (Chap. 10).

Applying the steps above it is possible to realize sophisticated model-based control strategies for trajectory tracking in case of complex parallel robots. This has been demonstrated exemplarily and successfully for the planar parallel robot SpiderMill in this thesis.

F Summary in German

Die in der industriellen Praxis am weitesten verbreiteten Roboter sind von serieller Struktur. Sie sind hoch flexibel und kosteneffizient. Seit Jahren stellen sie den industriellen Standard dar. Jedoch ist eine weitergehende Steigerung ihrer Leistungsfähigkeit aufgrund ihres strukturellen Aufbaus begrenzt und i.d.R. bereits ausgeschöpft. Im Fall von seriellen Strukturen akkumulieren sich die durch die einzelnen Achsen verursachten Ungenauigkeiten und flexiblen Effekte [Abd07]. Dieser den Manipulatoren dieser Strukturklasse inhärente Nachteil kann nicht überwunden werden. Eine höhere Steifigkeit führt typischerweise zu einer höheren bewegten Maschinenmasse und reduziert damit die erreichbare Verfahrgeschwindigkeit (bei Handhabungsanwendungen) bzw. Zykluszeit (bei der Werkstückbearbeitung). Aus diesem Grund nimmt der Einsatz von Parallelrobotern in zeitlich sensitiven, hochdynamischen Anwendungen sowie bei Prozessen, die eine hohe strukturelle Steifigkeit eines Manipulators erfordern, stetig zu. Jedoch weist auch diese Klasse von Roboterstrukturen gewisse Nachteile auf. Die wichtigsten hiervon sind ein ungünstiges Verhältnis von Arbeits- zu Bauraum sowie ein in der Regel erhöhter Modellierungs- und Regelungsaufwand. Insbesondere die echtzeitfähige Realisierung effizienter Verfahren zu Trajektorienfolgeregelung stellt ein großes Problem dar.

Bis heute werden die meisten Parallelroboter immer noch gesteuert oder auf Basis stark vereinfachter dynamischer Modelle geregelt. Der Grund für den letztgenannten Punkt ist die im Allgemeinen sehr hohe Modellkomplexität in Verbindung mit dieser Klasse von Robotern. Dieser Aspekt wird weiter verschärft im Falle von komplexen Parallelrobotern. Unter diesen werden im Rahmen dieser Arbeit parallelkinematische Manipulatoren verstanden, die strukturelle oder funktionale Redundanzen aufweisen. Um jedoch auch im Falle dieser Klasse von Robotern fortschrittliche Verfahren zur modellbasierten Trajektorienfolgeregelung direkt umsetzen zu können, sind akkurate, kompakte und echtzeitfähig implementierbare dynamische Modelle erforderlich.

Jedoch stellen kompakte, dynamische Modelle lediglich einen ersten Schritt auf dem Weg zu einer modellbasierten Trajektorienfolgeregelung mit einer hohen Bahngenauigkeit und geringen Verfahrzeiten dar. Zusätzlich sind geeignet geplante Trajektorien erforderlich. Bei deren Planung ist es notwendig, die physikalischen Begrenzungen der Roboter und dabei insbesondere die ihrer Antriebssysteme mit zu berücksichtigen. Bei Parallelrobotern ist dies u.a. aufgrund ihrer mehreren Führungsketten eine anspruchsvolle Aufgabe, für die bisher nur wenige Basisansätze existieren.

Ziel dieser Arbeit ist die Entwicklung eines ganzheitlichen Ansatzes zur modellbasierten Trajektorienfolgeregelung von komplexen Parallelrobotern. Dabei stellen sowohl die Herleitung kompakter dynamischer Modelle als auch die Trajektorienplanung wesentliche Kernprobleme da, die im Rahmen dieser Arbeit zu lösen sind. Für die Realisierung der modellbasierten Regelungskonzepte sind jedoch noch weitere Vorarbeiten (Zwischenschritte) erforderlich. Beispielsweise erfordern fast alle von ihnen die Werte der Gelenkgeschwindigkeiten (bzw. Spindelgeschwindigkeiten), die über geeignete Beobachtungskonzepte ermittelt werden müssen. In dieser Arbeit werden im Rahmen der Zwischenschritte, wo immer möglich, Standardverfahren eingesetzt.

Letztendlich sind die Ergebnisse aller Schritte erforderlich, um modellbasierte Trajektorienfolgeregelungen für die SpiderMill realisieren zu können. Aufgrund der geleisteten Vorabeit können dabei Verfahren, wie die *Inverse Dynamics Control* direkt umgesetzt werden. Im Fall von komplexen Parallelrobotern für die, die dynamischen Modelle direkt hergeleitet werden – ohne eine vorhergehende Vereinfachung der Struktur – ist dieses Verfahren, aufgrund seiner hohen Rechenlast, i.d.R. nicht direkt implementierbar. Vielmehr kann es erst umgesetzt werden, nachdem das dynamische Modell auf seine dominanten Anteile reduziert wurde. Dadurch nimmt jedoch die Modellgenauigkeit ab und konsequenterweise die Trajektorienfolgefehler zu.

Zusammenfassend kann gesagt werden, dass in dieser Arbeit ein ganzheitlicher Ansatz zur modellbasierten Trajektorienfolgeregelung für komplexe Parallelroboter vorgestellt und am planaren Parallelroboter SpiderMill demonstriert wird. Dabei wird wo immer erforderlich auf die Besonderheiten des Demonstrators eingegangen.

Im Folgenden werden alle Schritte auf dem Weg zur modellbasierten Trajektorienfolgeregelung anhand der einzelnen Kapitel bzw. deren Ergebnisse näher erläutert. Die einzelnen Kapitel werden dabei zueinander und zum Gesamtkontext, wo immer sinnvoll möglich, in Bezug gesetzt.

Kapitel 2:

In diesem Kapitel werden zunächst die Vor- und Nachteile von seriell- und parallelkinematischen Robotern ganz allgemein erarbeitet. Anhand dieser Gegenüberstellung zeichnet sich bereits ab, welche Punkte bei der hier zu untersuchenden Klasse von Robotern Schwierigkeiten bereiten werden. Anschließend wird die am *Lehrstuhl für Rechneranwendungen in der Konstruktion* der Technischen Universität Kaiserslautern entwickelte und gefertigte Parallelkinematik SpiderMill vorgestellt. Es handelt sich bei ihr um eine aus möglichst vielen Standard- und Wiederholbauteilen aufgebaute parallelkinematische Struktur. Sie verfügt mit ihren drehbar gelagerten und mit der Struktur über Kugelumlaufspindeln verbundenen Motoren über ein für Parallelkinematiken recht ungewöhnliches Antriebskonzept. Der Manipulator ist als Fräsroboter für das Rapid Prototyping von Musterbauteilen aus weichen Materialien, wie beispielsweise Kunstharzen, konzipiert.

Ferner werden in diesem Kapitel die zur simulativen Untersuchung bzw. Umsetzung der Konzepte erstellten Simulationsmodelle eingeführt. Dabei handelt es sich um ein am *Lehrstuhl für Maschinenelemente und Getriebetechnik* im Mehrköpersimulationsprogramm MSC.ADAMS erstelltes Modell sowie ein weiteres für die Implementierung von Regelungskonzepten besser geeignetes in SimMechanics (Matlab/Simulink).

Darüber hinaus werden die Sensorik und Aktuatorik der Struktur sowie die Umgebung zur experimentellen Erprobung vorgestellt. Bei letzterer handelt es sich um die Anbindung des Demonstrators an ein dSPACE-System (DS1104 R&D Controller Board). Das in diesem Zusammenhang vorgestellte Antriebsmodell wird sowohl als Teil der Simulationsumgebung als auch in Kapitel 10 für den Entwurf einer Einzelachsregelung verwendet.

Durch dieses Kapitel ist die Ausgangssituation umrissen. Ferner sind sowohl die simulative als auch die experimentelle Testumgebung eingeführt.

Kapitel 3:

In Kapitel 3 wird eine strukturelle Analyse der Parallelkinematik SpiderMill durchgeführt. Diese umfasst die Bestimmung ihres Freiheitsgrades auf Basis eines erweiterten Grübler-Kutzbach Kriteriums sowie die Herleitung ihrer direkten und inversen Kinematik. Im Falle von Parallelrobotern ist, im Gegensatz zu seriellen Manipulatoren, die Bestimmung der inversen Kinematik „einfach" und die Lösung der direkten in der Regel nur numerisch möglich. Bei der SpiderMill können jedoch beide Beschreibungen analytisch hergeleitet werden. Die Gleichungen werden anschließend mit Hilfe des steifen MSC.ADAMS-Modells des Roboters verifiziert.

Einen weiteren wichtigen Punkt stellt die Arbeitsraumanalyse dar. Ohne sie und die sich anschließende Singularitätsanalyse ist eine sinnvolle Bahnplanung (Kapitel 7) für einen Roboter nicht möglich. Da die bewegliche Plattform der SpiderMill nicht umorientiert werden kann, handelt es sich bei dem ermittelnden Arbeitsraum um einen translatorischen. Für seine Bestimmung werden neben den Begrenzungen der aktiven bzw. pseudo-aktiven Gelenke auch die der passiven innerhalb der Führungsketten sowie auftretende Kontaktsituationen mit berücksichtigt. Im Gegensatz zu gängigen numerischen Methoden, die den Arbeitsraum lediglich rastern, werden hier die Arbeitsraumgrenzen auf Basis einer geometrischen Analyse bestimmt. Durch die Ermittlung von Streckkonfigurationen erfolgt ferner eine Zweiteilung des Arbeitsraums in einen nutzbaren unteren sowie einen lediglich theoretisch nutzbaren oberen. Diese sind durch eine Aneinanderreihung singulärer Konfigurationen (entsprechen den Streckkonfigurationen) voneinander getrennt.

Ein weiterer auf der Kinematik bzw. der differentiellen Kinematik basierender Punkt ist die Singularitätsanalyse. Diese wird für die drei bei der SpiderMill potentiell vorkommenden Klassen von Singularitäten durchgeführt. Dabei können direkte und inverse kinematische Singularitäten ermittelt und geometrisch interpretiert werden. Aufgrund dieser geometrischen Interpretation kann das Vorkommen kombinierter Singularitäten ausgeschlossen werden.

Die in diesem Kapitel durchgeführte Analyse ist relativ spezifisch für die Spider-Mill. Lediglich die bei der Analyse des Arbeitsraums berücksichtigten Punkte sind in dieser Form auch auf andere Parallelroboter übertragbar.

Kapitel 4:

In diesem Kapitel wird eine neue Strategie zur Bestimmung kompakter analytischer Modelle für komplexe Parallelroboter eingeführt. Prinzipiell sind zwei Wege zur Erreichung dieses Ziels möglich. Entweder wird ein dynamisches Modell für eine bestehende Struktur einfach hergeleitet und anschließend auf Basis einer Modellreduktionsstrategie (wie beispielsweise balanciertes Abschneiden) reduziert oder die Struktur wird vor der dynamischen Modellbildung vereinfacht. Hier wird der zweite Weg beschritten. Basierend auf dem Konzept der dynamischen äquivalenten Masse wird ein schrittweises Verfahren zur systematischen Vereinfachung redundanter parallelkinematischer Strukturen eingeführt. In diesem Zusammenhang werden auch potentielle Quellen für spätere Modellabweichungen analysiert und bewertet. Im Anschluss daran wird die Strategie auf die Parallelkinematik SpiderMill unter Berücksichtigung ihrer strukturellen Besonderheiten, wie des Antriebskonzepts, angewendet. Die so erhaltene vereinfachte Struktur bildet die Grundlage für eine dynamische Modellbildung im nachfolgenden Kapitel. Aufgrund ihrer deutlich geringeren Anzahl an Gliedern und Gelenken kann dabei ein deutlich kompakteres dynamisches Modell erzielt werden.

Die eingeführte Strategie ist nicht auf die SpiderMill beschränkt und kann in dieser Form auch auf andere Roboter der Klasse angewendet werden.

Kapitel 5:

In diesem Kapitel werden Standardverfahren zur dynamischen Modellbildung (Lagrange, Newton-Euler und virtuelle Leistung) auf die vereinfachte Struktur aus Kapitel 4 angewendet und analytische dynamische Modelle hergeleitet. Diese werden anschließend anhand des SimMechanics-Models der SpiderMill verifiziert. Alle hergeleiteten Modelle sind dabei für eine weitergehende Verwendung geeignet. Neben den Modellen für die vereinfachte Struktur wird auch eines für die fast vollständige (nur sehr partielle Anwendung der Modellvereinfachungsstrategie) hergeleitet. Anhand dessen tiefergehender Analyse kann die Berechtigung einiger der Schritte aus Kapitel 4 nachgewiesen werden. Nachdem die Modelle auf eine einheitliche Form gebracht wurden, wird deren Komplexität miteinander verglichen. In Abhängigkeit vom dynamischen Ansatz werden unterschiedlich kompakte und somit effiziente Modelle erzielt. Die mit der Strukturvereinfachung einhergehende Reduktion der Modellkomplexität ist dabei deutlich zu erkennen.

Neben den vollständigen dynamischen Modellen, die für eine modellbasierte Trajektorienfolgeregelung (Kapitel 10) erforderlich sind, werden auch drei vereinfachte Modelle hergeleitet. Die ersten beiden – ein Modell des linken Parallelkurbelgetriebes sowie eines der linken Führungskette inklusive der beweglichen Plattform – dienen dabei einer weiteren Untermauerung der Modellvereinfachungsstrategie. Das zusätzlich hergeleitete Einzelachsmodell dient als Grundlage für die spätere Realisierung eines nicht-modellbasierten dezentralisierten Regelungskonzepts (Kapitel 10).

Ferner wird auf die Möglichkeiten der Simulation mechatronischer Systeme eingegangen. Dabei werden zunächst die theoretischen Grundlagen der beiden Mehrkörpersimulationssysteme MSC.ADAMS und SimMechanics kurz vorgestellt. Anschließend werden die beiden in den Programmen implementierten Modelle der SpiderMill miteinander verglichen, mit dem Resultat, dass beide für die Analyse der dynamischen Zusammenhänge eingesetzt werden können. Das Fazit einer kurzen Analyse des Aspekts der Co-Simulation zeigt jedoch, dass lediglich das SimMechanics-Modell einen sinnvollen Entwurf von Regelungsalgorithmen erlaubt. Daher wird es auch zur Verifikation dieser Algorithmen im Folgenden eingesetzt. Das MSC.ADAMS Model als ein Standard im Bereich der Mehrkörpersimulation kommt jedoch für den Test der Identifikationsstrategie im nachfolgenden Kapitel zum Einsatz.

Die in diesem Kapitel eingesetzten Verfahren zur dynamischen Modellbildung sind etablierte Standardverfahren. Die mit ihnen erzielten Ergebnisse in Verbindung mit der Vereinfachungsstrategie (Kapitel 4) untermauern deren Ansatz und zeigen auch ihre Unabhängigkeit vom nachgelagerten Modellierungskonzept.

Kapitel 6:

Da die Verwendung von modellbasierten Regelungskonzepten neben einem möglicht kompakten auch ein möglichst exaktes dynamisches Modell erforderlich macht, ist eine Identifikation der dynamischen Parameter zur Steigerung der Modellgenauigkeit erforderlich. Insbesondere die experimentelle Identifikation von Reibparametern am physikalischen Demonstrator ist unumgänglich.

Das hier eingesetzte Verfahren ist eine Kombination zweier in der Literatur bekannter Verfahren der direkten und der indirekten Identifikation. Dabei wird der Tatsache Rechnung getragen, dass bei einer Parallelkinematik auf Grund der verketten Struktur eine Identifikation der einzelnen Achsparameter nicht in einer sequentiellen Form, wie bei seriellen Kinematiken üblich, möglich ist. Ferner wird auch der Einfluss der passiven Gelenkreibung mit berücksichtigt.

Das so erhaltene dynamische Modell mit identifizierten Parametern bildet die Grundlage für die modellbasierten Regelungsverfahren (Kapitel 10), insbesondere wenn es um deren experimentelle Umsetzung geht. Im Rahmen seiner Verifikation konnte die erwartete, weitere Verbesserung der Modellgenauigkeit beobachtet werden.

Kapitel 7:

In Kapitel 7 wird ein neues Verfahren zur Planung glatter, zeitoptimaler Trajektorien im Phasenraum der Pfadvariable vorgestellt. Es erweitert hierzu bestehend Basisalgorithmen, die ebenfalls im Phasenraum bzw. den davon abgeleiteten Phasenebenen arbeiten. Durch eine Ergänzung der Nebenbedingungen des Optimierungsproblems werden alle wesentlichen physikalischen Beschränkungen eines Manipulators erfasst.

Um das komplexe Optimierungsproblem zu lösen, wird anschließend dessen Lösungsraum reduziert. Diese Vereinfachung ist bei den im Weiteren geplanten Referenzbahnen bzw. Trajektorien möglich. In einem zukünftigen Entwicklungsschritt ist jedoch noch eine Lösung des allgemeinen Problems zu erarbeiten.

Das entwickelte Verfahren basiert im Wesentlichen auf einer Glättung der maximalen Geschwindigkeitsgrenzkurve – die im untersuchten Fall eine erreichbare Kurve darstellt – sowie der Übergänge auf diese von Start- und Endpunkt. Durch die ausschließliche Verwendung der Pfadvariablen – und nicht der Zeit – als Integrationsvariable der eingeführten Modelle, ist eine anschauliche Interpretation des Verfahrens sowie dessen effektive Implementierbarkeit sichergestellt.

Basierend auf dem entwickelten Verfahren werden nach der Planung von Pfaden im Arbeitsraum die Referenztrajektorien für die Trajektorienfolgeregelung berechnet. Die eingeführte Strategie ist nicht auf die SpiderMill beschränkt und kann auch auf andere (Parallel-)Roboter angewendet werden.

Kapitel 8:

In diesem Kapitel wird das Prinzip der statischen Feedbacklinearisierung auf das Modell der Parallelkinematik SpiderMill angewendet. Es stellt eine in der Robotik typische Vorgehensweise zur Linearisierung der nichtlinearen Dynamik dar und bildet die Grundlage für einen linearen Reglerentwurf. In den meisten Fällen mündet es in der sogenannten *Inverse Dynamics Control*.

Für die eigentliche Feedbacklinearisierung sind zwei Schritte erforderlich. Die Transformation des Modells in einen neuen Zustandsraum, in dem es linearisierbar ist sowie der Entwurf einer das System linearisierenden Rückführung. Im Hinblick auf die *Inverse Dynamics Control* sind die aus der Feedbacklinearisierung resultierenden Doppelintegrator-Ketten in einem Folgeschritt durch eine geeignete Zustandsrückführung zu stabilisieren.

Bei der Anwendung des Prinzips auf die SpiderMill wird außerdem der Einfluss einer nicht exakten Ein-Ausgangs-Linearisierung näher analysiert. Ferner werden die Wirkungen verschiedener stabilisierender Verstärkungen für ein Testszenario simulativ untersucht.

Die Ergebnisse dieses Kapitels bilden die Grundlage für die Realisierung einiger modellbasierten Regelungskonzepte zur Trajektorienfolgeregelung (Kapitel 10). Im Falle dieser sind die Komponenten des dynamische Modells in Echtzeit zu rechnen. Aus diesem Grund kann die Feedbacklinearisierung in der Regel nicht direkt auf die Modelle komplexer Roboter angewendet werden. Durch die Schritte der vorangehenden Kapitel wird dies jedoch hier ermöglicht.

Kapitel 9:

Da sowohl für die Feedbacklinearisierung aus Kapitel 8 als auch für die meisten der Regelungskonzepte Informationen über die Gelenk- bzw. Spindelgeschwindigkeiten benötigt werden, werden in diesem Kapitel generelle Möglichkeiten für deren Bestimmung untersucht.

Für das feedbacklinearisierte System bieten sich Beobachtungskonzepte wie der High-Gain-Beobachter oder das diskrete lineare Kalman-Filter an. Beide werden unter Berücksichtigung einer nicht exakten Linearisierung näher analysiert. Ferner wird eine Klasse nicht modellbasierter Verfahren, das sogenannte *First-Order-Adaptive-Windowing* (FOAW) anhand, zweier seiner Vertreter – das *First-Order-* und das *Best-Fit FOAW* – simulativ untersucht.

Kapitel 10:

In Kapitel 10 werden verschiedene Verfahren zur Trajektorienfolgeregelung simulativ und experimentell miteinander verglichen. Ausgangspunkt bildet die *Inverse Dynamics Control*, die als Referenzstrategie eingeführt wird. Anhand ihr werden verschiedene Verstärkungen getestet sowie die unterschiedlichen Konzepte zur Bestimmung der Geschwindigkeiten miteinander verglichen. Ferner werden beobachtete Besonderheiten der simulativ und experimentell ermittelten Ergebnisse diskutiert.

Nach der Einführung und Analyse weiterer modellbasierter Regelungsverfahren, wie beispielsweise der PD-Regelung mit Schwerkraftkompensation, werden auch nicht-modellbasierte Verfahren untersucht. Dabei handelt es sich sowohl um eine zentralisierte (Zustandsregelung) als auch eine dezentralisierte (Einzelachsregelung) Strategie. Wie erwartet sind die modellbasierten Konzepte den nicht-modellbasierten deutlich überlegen.

Es bleibt hervorzuheben, dass die Anwendung der modellbasierten Verfahren ohne ein kompaktes, echtzeitfähig implementierbares Modell der Dynamik, wie es auf Grundlage der Modellvereinfachungsstrategie hergeleitet werden kann, in dieser Form nicht möglich wäre. Vielmehr müssten nicht dominante Anteile der Dynamik identifiziert und vernachlässigt werden, um die im Falle von redundanten Parallelmanipulatoren komplexen dynamischen Modelle auf einem System, wie dem hier verwendeten von dSPACE, mit einer geeigneten Abtastrate abarbeiten zu können.

Kapitel 11:

Da in jedem der vorangegangen Kapitel am Ende mögliche Erweiterungen der konkreten Einzelprobleme diskutiert werden, schließt sich an eine kurze Wiederholung aller erzielten Ergebnisse nur noch ein genereller Ausblick an. Der nächste wesentliche Schritt im Hinblick auf einen Rapid-Prototyping-Prozess ist die Erweiterung der Trajektorienfolgeregelung auf eine hybride Regelungsstrategie. Ferner sind im Hinblick auf die bei einer Werkstückbearbeitung auftretende Deformationen, Schwingungen sowie Krafteinleitungen in die Struktur Ansätze in Richtung einer flexiblen Modellbildung sowie aktiven Schwingungsbedämpfung näher zu untersuchen.

G Publications and supervised student thesis

During the time of this thesis the following publications have been made:

Machleidt, K., Kroneis, J., Liu, S.: *Stabilization of the Furuta Pendulum Using a Nonlinear Control Law Based on the Method of Controlled Lagrangians*, 2007 IEEE Int. Symp. on Industrial Electronics (ISIE'07), Vigo/Spain, 2007.

Kroneis, J., Gastauer, T., Liu, S., Sauer, B.: *Evaluierung einer Trajektorien-planungsstrategie höherer Ordnung für eine Parallelkinematik mit drei Freiheits-graden unter Verwendung einer Co-Simulation zwischen MSC.ADAMS und MAT-LAB/Simulink*, Kongress für Simulation im Produktentstehungsprozess (Sim-PEP), Würzburg/Germany, 2007.

Kroneis, J., Gastauer, T., Liu, S., Sauer, B.: *Flexible Body Modeling and Vibration Analysis of a Parallel Robot with Numerical and Analytical Methods for the Purpose of Active Vibration Damping*, 52. Internationales Wissenschaftliches Kolloquium (52nd IWK), Illmenau/Germany, 2007.

Kroneis, J., Liu, S.: *Hybrid Approach to Modeling a Flexible Parallel Robot for Vibration Damping*, 31st ASME Annual Mechanisms and Robotics Conference (IDETC/MECH'07), Las Vegas/USA, 2007.

Kroneis, J., Liu, S.: *Flexible Body Modeling and Vibration Damping for a Planar Parallel Robot Using Input Shaping*, 2007 IEEE/ASME Int. Conf. on Advanced Intelligent Mechatronics (AIM2007), Zurich/Switzerland, 2007.

Kroneis, J., Müller, P., Liu, S.: *Direct parameter identification for a complex parallel robot based on an analytically reduced model*, 17th IFAC World Congress (IFAC'08), Seoul/South Korea, 2008.

Kroneis, J., Müller, P., Liu, S.: *Reduced Order Modeling and Direct Parameter Identification for a Complex Parallel Robot*, 32nd ASME Annual Mechanisms and Robotics Conference (IDETC/MECH'08), New York/USA, 2008.

Görges, D., Kroneis, J., Liu, S.: *Active Vibration Control of Storage and Re-trieval Machines*, 32nd ASME Annual Mechanisms and Robotics Conference (IDETC/MECH'08), New York/USA, 2008.

Kroneis, J., Müller, P., Liu, S.: *Simplified Modelling and Parameter Identification of Parallel Robots with Redundancies* , Journals of Systems and Control Engineer-ing, Vol. 223, No. I1, February 2009.

The following student thesis have been supervised with Prof. Dr.-Ing. S. Liu and partly with other members of the research staff at the Institute of Control Systems:

Baumanm, M.: *Modelling and Verification of the Kinematics and Dynamics of the Parallel Manipulator Tripod*, 2006.

Demke, P.: *Entwurf und Implementierung von Regelungskonzepten für die Paral-lelkinematik SpiderMill mittels dSPACE*, 2008.

Ezeti Diang, J.: *Identifikation von Starrkörper- und Reibeinflüssen am Demon-strator SpiderMill*, 2010.

Feijoo, I.: *Dynamic Modelling of the parallel manipulator SpiderMill using Jour-dain's principle*, 2009.

Görges, D.: *Aktive Dämpfung der Mastschwingungen von Regalbediengeräten*, 2005.

Hengen, M.-Ph.: *Untersuchung der äquivalenten Modellierbarkeit des mobilen Roboters CARMA als Hybridstruktur*, 2008.

Kornow, J.: *Optimale Trajektorienplanung für eine parallelkinematische Struktur*, 2006.

Kornow, J.: *Analyse und aktive Bedämpfung von Schwingungen bei parallelkine-matischen Strukturen*, 2006.

Liu, L.: *Design and implementation of an open loop control concept for the parallel kinematic structure SpiderMill*, 2006.

Machleidt, K.: *Modellierung eines inversen Rotationspendels und Untersuchung nichtlinearer Regelungskonzepte zu seiner Stabilisierung*, 2005.

Mas, M.: *Full dynamic modelling of the parallel manipulator SpiderMill using Newton-Euler approach*, 2009.

Matzen, T.: *Flachheitsbasierte Regelung/Steuerung eines Linear-Pendels*, 2005.

Müller, P.: *Modellierung und Identifikation von Starrkörper- und Reibungsflüssen am Demonstrator Spidermill*, 2007.

Müller, P.: *Entwurf einer Folgeregelung entlang glatter, zeitoptimaler Trajektorien für die ebene Parallelkinematik SpiderMill*, 2008.

Prothmann, Ch.: *Analyse der Kinematik und Dynamik der parallelkinematischen Struktur Spidermill*, 2006.

Prothmann, Ch.: *Modellierung elastischer Effekte einer parallelkinematischen Struktur zur Modalanalyse und aktiven Schwingungsbedämpfung*, 2008.

Purba, W.: *Entwurf und Implementierung einer iterativ lernenden Positionsregelung für die Parallelkinematik SpiderMill*, 2009.

Simon, S.: *Entwurf und Implementierung von dezentralisierten und zentrallisierten Regelungskonzepten für die Parallelkinematik SpiderMill*, 2008.

Strohrmann, Ch.: *Modellierung elastischer Effekte in SimMechanics zur Modulanalyse und aktiven Schwingungsbedämpfung der Parallelkinematik SpiderMill*, 2009.

Tiras, F.: *Ausbau und Verbesserung des Steuerungs-/Regelungskonzepts einer parallelkinematischen Struktur*, 2006.

Zhao, J.: *Vergleich linearer und nichtlinearer Beobachtungskonzepte zur Schätzung der Geschwindigkeit der aktiven Gelenke des Demonstrators SpiderMill*, 2010.

H Curriculum vitae

Personal data

name:	Jens Kroneis
address:	Frühlingstr. 11
	67734 Sulzbachtal
date of birth:	10th June 1979
place of birth:	Kaiserslautern
marital status:	unmarried
nationality:	German

Education and military service

1985–1989	Grundschule Olsbrücken
1989–1995	Kurpfalz-Realschule Kaiserslautern
1995–1998	Technisches Gymnasium Kaiserslautern
07/1998–07/1999	Military service

Academic studies

1999–2004	Computer engineering at the University of Kaiserslautern, specializing in automation engineering
8th Nov. 2004	Diploma in Computer engineering (Informationstechnik)

Practical trainings and work experiences

08/1999–10/1999	Adam Opel AG, Kaiserslautern (basic internship)
07/2000–08/2000	Technische Werke Kaiserslautern (basic internship)
06/2002–03/2004	Student assistant, University of Kaiserslautern

| 11/2004–10/2009 | Scientific assistant at the Institute of Control Systems, University of Kaiserslautern |
| since 10/2010 | E&I engineer, BASF SE, Ludwigshafen |

11/2004–10/2009   Scientific assistant at the Institute of Control Systems, University of Kaiserslautern

01/2010–05/2010   Software engineer, Sirona Dental Systems GmbH, Bensheim

since 10/2010   E&I engineer, BASF SE, Ludwigshafen

Bibliography

[AAH06]   H. Abdellatif, H. Abdellatif, and B. Heimann. On compensation of passive joint friction in robotic manipulators: Modeling, detection and identification. In B. Heimann, editor, *Proc. IEEE Int. Conf. on Control Applications CCA '06*, pages 2510–2515, 2006.

[Abd07]   Houssem Abdellatif. *Modellierung, Identifikation und robuste Regelung von Robotern mit parallelkinematischen Strukturen*. PhD thesis, Institut für Robotik, Leibniz Universität Hannover, 2007.

[ABHG04]   H. Abdellatif, F. Benimelli, B. Heimann, and M. Grotjahn. Direct identification of dynamic parameters for parallel manipulators. *Proc. of the Int. Conf. on Mechatronics and Robotics, Aachen Germany*, pages 999–1005, 2004.

[ACC99]   G. Antonelli, F. Caccavale, and P. Chiacchio. A systematic procedure for the identification of dynamic parameters of robot manipulators. *Robotica*, 17(4):427–435, 1999.

[AGH05]   H. Abdellatif, M. Grotjahn, and B. Heimann. High efficient dynamics calculation approach for computed-force control of robots with parallel structures. In *Decision and Control, 2005 and 2005 European Control Conf.. CDC-ECC '05. 44th IEEE Conf. on*, pages 2024–2029, Dec. 2005.

[Agr91]   S.K. Agrawal. Inertia matrix singularity of planar series-chain manipulators. In *Robotics and Automation, 1991. Proc., 1991 IEEE Int. Conf. on*, pages 102–107 vol.1, Apr. 1991.

[Agr93]   Sunil K. Agrawal. Inertia matrix singularity of series-chain spatial manipulators with point masses. *J. of Dynamic Systems, Measurement, and Control*, 115(4):723–725, 1993.

[AH05]   H. Abdellatif and B. Heimann. Adapted time-optimal trajectory planning for parallel manipulators with full dynamic modelling. In *Int. Conf. on Robotics and Automation*, pages 413 – 418, 2005.

[BD67]   J. Bellantoni and K. Dodge. A square root formulation of the kalman-schmidt filter. *AIAA Journal*, 5:1309–1314, 1967.

[BDG85]   J. Bobrow, S. Dubowsky, and J. Gibson. Time-optimal control of robotic manipulators along specified paths. *Int. J. Robot. Res.*, 4(3):3–17, 1985.

[BDHZ98]   P.R. Belanger, P. Dobrovolny, A. Helmy, and X. Zhang. Estimation of Angular Velocity and Acceleration from Shaft-Encoder Measurements. *Int. J. of Robot. Res.*, 17(11):1225–1233, 1998.

[BH98]   G. Besançon and H. Hammouri. On observer design for interconnected systems. *Journal of Mathematical Systems, Estimation, and Control*, 8(3):1–25, 1998.

[BI05]   B. Bona and M. Indri. Friction compensation in robotics: an overview. In M. Indri, editor, *Proc. and 2005 European Control Conf. Decision and Control CDC-ECC '05. 44th IEEE Conf. on*, Seville, Spain, pages 4360–4367, 2005.

[BIG74]   A. Ben-Israel and T. N. Greville. *Generalized Inverses: Theory and Applications*. Wiley-Interscience, 1974.

[BKL+08]   Ch. Budde, M. Kolbus, Ph. Last, A. Raatz, J. Hesselbach, and W. Schumacher. Optimized Change of Working and Assemby Mode of the SFB 562 TRIGLIDE-Robot. In *3rd Int. Colloquium of the SFB 562: RSHA 2008*, pages 221–236, Aachen, 2008. Shaker Verlag.

[BLH07]   Ch. Budde, Ph. Last, and J. Hesselbach. Development of a Triglide-Robot with Enlarged Workspace. In *ICRA*, Rom, Italy, 2007.

[Bon02]   Ilian Bonev. Kinematics terminology related to parallel mechanisms. http://www.parallemic.org/Terminology/Kinematics.html, 2002.

[BZG03]   Ilian A. Bonev, Dimiter Zlatanov, and Clément M. Gosselin. Singularity analysis of 3-dof planar parallel mechanisms via screw theory. *J. Mech. Des.*, 125(3):573–581, Sept. 2003.

[BZH02]   G. Besançon, Q. Zhang, and H. Hammouri. High-gain observer based state and parameter estimation in nonlinear systems. In *Proc. of the 15th IFAC World Congress*, Barcelona, Spain, Jul. 2002.

[CA91]   K. Cleary and T. Arai. A prototype parallel manipulator: Kinematics, construction, software, workspace results, and singularity analysis. *Proc. of ICRA, Sacramento, California*, pages 566–571, Apr. 1991.

[CB97]   A. Codourey and E. Burdet. A body-oriented method for finding a linear form of the dynamic equation of fully parallel robots. In *Proc.*

IEEE Int. Conf. on Robotics and Automation, volume 2, pages 1612–1618 vol.2, 1997.

[CC00]   D. Constantinescu and E. A. Croft. Smooth and time-optimal trajectory planning for industrial manipulators along specified paths. *J. of Robotic Systems*, 17(5):233–249, 2000.

[Cha98]   D. Chablat. *Domain d'unicité et parcourabilité pour les manipulateurs pleinement parallèles*. PhD thesis, Ecole Centrale, Nantes, 1998.

[CLVD06]   Y. Chahlaoui, D. Lemonnier, A. Vandendorpe, and P. Van Dooren. Second-order balanced truncation. *Linear Algebra and its Applications - Special Issue on Order Reduction of Large-Scale Systems*, 415(2-3):373–384, 2006.

[Cod98]   A. Codourey. Dynamic modeling of parallel robots for computed-torque control implementation. *Int. J. of Robot. Res.*, 17(12):1325–1336, 1998.

[Con98]   Daniela Constantinescu. Smooth time optimal trajectory planning for industrial manipulators. Master's thesis, Department of Mechanical Engineering, The University of British Columbia, 1998.

[Cra05]   John J. Craig. *Introduction to Robotics - Mechanics and Control*. Pearson Education, Inc., Upper Saddle River, NJ, 3. edition, 2005.

[Cro00]   Nico Croon. Entwicklung einer modularen Fräsmaschine, Diplomarbeit, Fachbereich Maschinenbau, Lehrstuhl für virtuelle Produktentwicklung, University of Kaiserslautern, Apr. 2000.

[DA98]   M. Daemi-Avval. *Modellierung und Identifikation der Dynamik von Industrierobotern für den Einsatz in Regelungen*. PhD thesis, Mechatronik-Zentrum Hannover, Leibniz Universität Hannover, 1998.

[Dah92]   O. Dahl. *Path constrained robot control*. PhD thesis, Department of Automatic Control, Lund Institute of Technology, Lund, 1992.

[Dau01]   Olaf Dauber. *Elastohydrodynamische Rollreibung in Stahl-Keramik-Kontakten*. PhD thesis, Fakultät für Maschinenbau der Universität Karlsruhe (TH), 2001.

[DB05]   Richard C. Dorf and Robert H. Bishop. *Modern control systems*. Pearson Prentice Hall, Upper Saddle River, NJ, 10. ed., internat. ed. edition, 2005.

[Dem08]   P. Demke. Entwurf und Implementierung von Regelungskonzepten für die Parallelkinematik SpiderMill mittels dSPACE, Studienarbeit, Institute of Control Systems, University of Kaiserslautern, 2008.

[DFM05]   Bodmar Diestel-Feddersen and Giulio Milighetti. Strukturvariable Regelung eines humanoiden Roboterarmes mit bildgebenden und Kraft-Momenten-Sensoren. In *36. Sitzung GMA-FA 4.13*, Gersthofen, Germany, Jan. 2005.

[DHL06]   J. S. Dai, Z. Huang, and H. Lipkin. Mobility of Overconstrained Parallel Mechanisms. *J. of Mechanical Design*, 128(1):220–229, 2006.

[Diz66]   Bekir Dizioğlu. *Getriebelehre - Band 3 - Dynamik*. Friedr. Vieweg & Sohn GmbH, Braunschweig, 1966.

[DK97]   A. Dabroom and H.K. Khalil. Numerical differentiation using high-gain observers. In *Decision and Control, 1997., Proc. of the 36th IEEE Conference on*, volume 5, pages 4790–4795, Dec. 10-12 1997.

[Dor97]   Richard C. Dorf. *The electrical engineering handbook*. CRC Press, Boca Raton, 2. edition, 1997.

[DRMFP08]   Miguel Díaz-Rodríguez, Vicente Mata, Nidal Farhat, and Sebastian Provenzano. Identifiability of the dynamic parameters of a class of parallel robots in the presence of measurement noise and modeling discrepancy. *Mechanics Based Design of Structures and Machines*, 36:478–498, 2008.

[DW06]   A. Degani and A. Wolf. Graphical singularity analysis of planar parallel manipulators. In *Proc. IEEE Int. Conf. on Robotics and Automation ICRA 2006*, pages 751–756, 2006.

[dWL97]   C. Canudas de Wit and P. Lischinsky. Adaptive friction compensation with partially known dynamic friction model. *Int. J. of Adaptive Control and Signal Processing*, 11(1):65–80, 1997.

[ED10]   J. Ezeti Diang. Identifikation von Starrkörper- und Reibeinflüssen am Demonstrator SpiderMill, Studienarbeit, Institute of Control Systems, University of Kaiserslautern, 2010.

[EHE08]   Ayssam Y. Elkady, Sarwat N. Hanna, and Galal A. Elkobrosy. On the modeling and control of the cartesian parallel manipulator. *Advances in Computer and Information Sciences and Engineering*, pages 90–96, 2008.

[ER94]   R. E. Ellis and S. L. Ricker. Two numerical issues in simulating constrained robot dynamics. *IEEE Trans. Syst., Man, Cyber*, 24(1):19–27, 1994.

[ES96]   A. G. Erdman and G. N. Sandor. *Mechanism Design: Analysis and Synthesis: Vol. 1.* Prentice Hall College Div, 3 edition, Nov. 1996.

[F94]   O. Föllinger. *Regelungstechnik: Eine Einführung in die Methoden und ihre Anwendung.* Hüthig Buch Verlag Heidelberg, 8. edition, 1994.

[FCSB00]   A. Frisoli, D. Checcacci, F. Salsedo, and M. Bergamasco. Synthesis by screw algebra of translating in-parallel actuated mechanisms. *Advances in Robot Kinematics*, pages 433–440, 2000.

[Fei09]   I. Feijoo. Dynamic Modelling of the parallel manipulator Spider-Mill using Jourdain's principle, Diplomarbeit, Institute of Control Systems, University of Kaiserslautern, 2009.

[FMAPV08]   Nidal Farhata, Vicente Mataa, Álvaro Pageb, and Francisco Valero. Identification of dynamic parameters of a 3-dof rps parallel manipulator. *Mechanism and Machine Theory*, 43(1):1–17, Jan. 2008.

[FMS95]   C. Ferraresi, G. Montaccini, and M. Sorli. Workspace and dexterity evaluation of 6 d.o.f. spatial mechanism. *Proc. of the 9th World Congress on the Theory of Machines and Mechanisms, Milan*, pages 57–61, August 1995.

[FPEN06]   Gene F. Franklin, David J. Powell, and Abbas Emami-Naeini. *Feedback Control of Dynamic Systems.* Prentice Hall PTR, Pearson Prentice Hall, Upper Saddle River, NJ, USA, 5 edition, 2006.

[FPW90]   Gene F. Franklin, David J. Powell, and Michael L. Workman. *Digital Control of Dynamic Systems.* Addison-Wesley Publishing Company, Inc., 2 edition, 1990.

[GA90]   C. Gosselin and J. Angeles. Singularity analysis of closed-loop kinematic chains. *IEEE Trans. Robot. Autom.*, 6(3):281–290, 1990.

[GAH04]   M. Grotjahn, H. Abdellatif, and B. Heimann. Path accuracy improvement of parallel kinematic structures by the identification of friction and rigid-body dynamics. In *The 4th Chemnitz Parallel Kinematics Seminar PKS 2004*, pages 217–236, Chemnitz, Germany, 2004.

[Gau90]   M. Gautier. Numerical calculation of the base inertial parameters of robots. *IEEE Int. Conf. on Robotics and Automation*, pages 1020–1025, 1990.

[GH00]   M. Grotjahn and B. Heimann. Symbolic calculation of robots' base reaction-force/torque equations with minimal parameter set. *13th*

CISM-IFToMM Symposium on the Theory and Practise of Robots and Manipulators (RoManSy), pages 59–65, 2000.

[GHA04a]   M. Grotjahn, B. Heimann, and H. Abdellatif. Identification of friction and rigid-body dynamics of parallel kinematic structures for model-based control. *Multibody System Dynamics*, 11(3):273–294, 2004.

[GHA04b]   Martin Grotjahn, Bodo Heimann, and Houssem Abdellatif. Identifcation of friction and rigid-body dynamics of parallel kinematic structures for model-based control. *Multibody System Dynamics*, 11(3):273–294, 2004.

[GHO92]   J.P. Gauthier, H. Hammouri, and S. Othman. A simple observer for nonlinear systems applications to bioreactors. *Automatic Control, IEEE Transactions on*, 37(6):875–880, Jun. 1992.

[GJKHG02] M. Grotjahn, J. J. Kühn, B. Heimann, and H. Grendel. Dynamic equations of parallel robots in minimal dimensional parameter-linear form. In *Proc. of the 14th CISM-IFToMM Symposium on the Theory and Practice of Robots and Manipulators (RoManSy)*, pages 67–76, 2002.

[GK86]   M. Gautier and W. Khalil. A direct determination of minimum inertial parameters of robots. *IEEE Int. Conf. on Robotics and Automation*, pages 1682–1687, 1986.

[GK90]   M. Gautier and W. Khalil. Direct calculation of minimum set of inertial parameters of serial robots. *IEEE Trans. on Robotics and Automation*, pages 368–373, 1990.

[GMS05]   P. E. Gill, W. Murray, and M. A. Saunders. SNOPT: An SQP algorithm for large-scale constrained optimization. In *SIAM Review*, volume 47, pages 99–131, 2005.

[GPS02]   H. Goldstein, C. Poole, and J. Safko. *Classical Mechanics - Third Edition*. Addison Wesley Longman, 3. edition, 2002.

[Gro03]   Martin Grotjahn. *Kompensation nichtlinearer dynamischer Effekte bei seriellen und parallelen Robotern zur Erhöhung der Bahngenauigkeit*. PhD thesis, Mechatronik-Zentrum Hannover - Leibniz Universität Hannover, May 2003.

[GST+05]   D. T. Griffith, A. J. Sinclair, J. D. Turner, J. E. Hurtado, and J. L. Junkins. *Spaceflight Mechanics 2004 (Advances in the Astronautical Sciences)*, chapter Automatic Generation and Integration of Equations of Motion by Operator Over-Loading Techniques (AAS 04-242), pages 2167–2188. American Astronautical Society, 2005.

[GVCW04] C.M. Gosselin, F. Vollmer, G. Cote, and Yangnian Wu. Synthesis and design of reactionless three-degree-of-freedom parallel mechanisms. *IEEE Trans. on Robotics and Automation*, 20(2):191–199, 2004.

[Hö05] R. Höpler. *A Unifying Object-oriented Methodology to Consolidate Multibody Dynamics Computations in Robot Control*, volume 1054, pages 1 – 115. PhD Thesis, VDI Verlag, 2005.

[Hag06] Peter Hagedorn. *Technische Mechanik*, volume 3. Verlag Harri Deutsch, 2006.

[HCMZ06] T. Huang, D.G. Chetwynd, J.P. Mei, and X.M. Zhao. Tolerance design of a 2-dof overconstrained translational parallel robot. *IEEE Trans. on Robotics*, 22(1):167–172, 2006.

[HD00] Peter Häse and Christoph Decking. Investigation of drive systems using adams and matlab/simulink. In *ADAMS/RAil Users' Conference*, Harlem, Netherlands, 2000.

[HJSC97] Vincent Hayward, Farrokh Janabi-Sharifi, and Chung-Shin Jason Chen. Adaptive windowing discrete-time velocity estimation techniques: Application to haptic interfaces. In *In Rob. Control. SY.RO.CO'97*, IFAC, 1997.

[HL03] Z. Huang and Q. C. Li. Type synthesis of symmetrical lower-mobility parallel mechanisms using the constraint-synthesis method. *The Int. J. of Robotics Research*, 22(1):59–79, Jan. 2003.

[HLL08] Z. Huang, J. F. Liu, and Q. C. Li. Unified methodology for mobility analysis based on screw theory. In L. Wand and J. Xi, editors, *Smart Devices and Machines for Advanced Manufacturing*, pages 49–78, 2008.

[HM98] K. H. Hunt and P. R. McAree. The octahedral manipulator: geometry and mobility. In *Int. J. Rob. Res.*, volume 17, pages 868–885, 1998.

[Hoc70] B. A. Hockey. The method of dynamically similar systems applied to the distribution of mass in spatial mechanisms. *J. of Mechanisms*, 5:169–180, 1970.

[Hol80] John M. Hollerbach. A recursive lagrangian formulation of manipulator dynamics and a comparative study of dynamics formulation complexity. *Systems, Man and Cybernetics, IEEE Trans. on*, 10(11):730–

736, Nov. 1980.

[HS91]   J.M. Hervé and F. Sparacino. Structural synthesis of 'parallel' robots
           generating spatial translation. In *Proc. Fifth Int. Conf. on Advanced
           Robotics 'Robots in Unstructured Environments', 91 ICAR, Pisa,
           Italy*, volume 1, pages 808–813, 1991.

[HS92]   N. M Hervé and F. Sparacino. Star, a new concept in robotics. In *3rd
           Int. Workshop on Advances in Robotics Kinematics, Ferrara, Italy*,
           pages 176–183. Sept. 7-9 1992.

[HS04]   H. S. Han and J. H. Seo. Design of a multi-body dynamics anal-
           ysis program using the object-oriented concept. *Adv. Eng. Softw.*,
           35(2):95–103, 2004.

[HSvS05]   Robert Höpler, Maximilian Stelzer, and Oskar von Stryk. Object-
           oriented dynamics modeling for simulation, optimization and control
           of walking robots. *Proc. 18th Symposium on SImulation Technique,
           ASIM*, pages 588–593, Sept. 12-15 2005.

[Hua04]   Z. Huang. The kinematics and type synthesis of lower-mobility par-
           allel robot manipulators. *Proc. of the 11 th World Congress in Mech-
           anism and Machine Science, Tianjin, China*, pages 65–76, 2004.

[Hun90]   K. H. Hunt. *Kinematic geometry of mechanisms -The Oxford engi-
           neering science series - 7*. Oxford University Press, first published
           1978, reprinted in paperback (with corrections) edition, 1990.

[IPC92]   C. Innocenti and V. Parenti-Castelli. Singularity-free evolution from
           one configuration to another in serial and fully-parallel manipulators.
           In *22nd Biennial Mechanisms Conf.*, pages 553–560, Sept. 1992.

[Isi89]   Alberto Isidori. *Nonlinear Control Systems: an introduction*.
           Springer Verlag, 2. edition, 1989.

[JH02]   Harold Josephs and Ronald L. Huston. *Dynamics of Mechanical
           Systems*. CRC Press, 2002.

[JK97]   S. J. Julier and J. K.Uhlmann. A New Extension of the Kalman
           Filter to Nonlinear Systems. In *Proc. of Areosense: 11th Int. Sympos.
           Aerospace/Defense Sensing, Simulation, and Controls*, Orlando, FL,
           1997.

[Jou09]   P. F. B. Jourdain. Note on an analogue of gauss principle of least
           constraint. *Quarterly J. of Pure and Applied Mathematics*, 40:153–
           157, 1909.

[JSHC00]   F. Janabi-Sharifi, V. Hayward, and C.-S.J. Chen. Discrete-time adaptive windowing for velocity estimation. *Control Systems Technology, IEEE Transactions on*, 8(6):1003–1009, Nov. 2000.

[JU04]   S.J. Julier and J.K. Uhlmann. Unscented filtering and nonlinear estimation. *Proceedings of the IEEE*, 92(3):401–422, Mar. 2004.

[JYAXLL01] Q. Jin, T. L. Yang, F. H. Yaho A. X. Liu, and Y. F. Luo. Structural synthesis and classification of the 4dof(3t-1r) parallel robot mechanisms based on the units of a single-opened-chain. *China Mechanical Engineering*, 12(9):1038–1041, 2001.

[Kan61]   T. R. Kane. Dynamics of nonholonomic systems. *ASME J. of Applied Mechanics*, 28:574–578, 1961.

[KD02]   W. Khalil and E. Dombre. *Modeling, Identification & Control of Robots*. Hermes Penton Science, 2002.

[Kel97]   Rafael Kelly. PD Control with Desired Gravity Compensation of Robotic Manipulators: A Review. *The Int. J. of Robotics Research*, 16(5):660–672, 1997.

[Ker94]   H. Kerle. Parallelroboter in der Handhabungstechnik - Bauformen, Berechnungsverfahren. *VDI-Berichte*, 1111:207–227, 1994.

[KES04]   Kris Kozak, Imme Ebert-Uphoff, and William Singhose. Locally linearized dynamic analysis of parallel manipulators and application of input shaping to reduce vibrations. *J. of Mechanical Design*, 126:156–168, 2004.

[KG85]   W. Khalil and M. Gautier. On the derivation of the dynamic models of robots. *Proc. ICAR, Tokyo, Japan*, pages 243–250, 1985.

[KG02]   Xianwen Kong and Clément M. Gosselin. Type synthesis of 3-dof spherical parallel manipulators based on screw theory. *ASME 2002 Int. Design Engineering Technical Conferences and Computers and Information in Engineering Conf. (IDETC/CIE2002), Montreal, Quebec, Canada*, 5: 27th Biennial Mechanisms and Robotics Conf.:851–856, 2002.

[KGLS07a] J. Kroneis, T. Gastauer, S. Liu, and B. Sauer. Evaluierung einer Trajektorienplanungsstrategie höherer Ordnung für eine Parallelkinematik mit drei Freiheitsgraden unter Verwendung einer Co-Simulation zwischen MSC.ADAMS und MATLAB/Simulink. Würzburg, Germany, 2007.

[KGLS07b] J. Kroneis, T. Gastauer, S. Liu, and B. Sauer. Flexible Body Modeling and Vibration Analysis of a Parallel Robot with Numerical and

Analytical Methods for the Purpose of Active Vibration Damping. *52. Internationales Wissenschaftliches Kolloquium (52nd IWK), Illmenau/Germany*, 2007.

[KH02]   M. Karouia and J. M. Hervé. A famliy of novel orientation 3-dof parallel robots. *14th RoManSy*, pages 359–368, Jul. 2002.

[Kha02]   H. K. Khalil. *Nonlinear Systems*. Prentice-Hall, Inc., 3. edition, 2002.

[KK86]   W. Khalil and J. Kleinfinger. A new geometric notation for open and closed-loop robots. In *Proc. IEEE Int. Conf. on Robotics and Automation*, volume 3, pages 1174–1179, San Francisco, 1986.

[KK87]   W. Khalil and J.-F. Kleinfinger. Minimum operations and minimum parameters of the dynamic models of tree structure robots. *Robotics and Automation, IEEE J. of*, 3(6):517–526, Dec. 1987.

[KL85]   Thomas R. Kane and David A. Levinson. *DYNAMICS: Theory and Applications (McGraw-Hill Series in Mechanical Engineering)*. McGraw-Hill, Inc., 1985.

[KML08a]   J. Kroneis, P. Müller, and S. Liu. Direct parameter identification for a complex parallel robot based on an analytically reduced model. In *17th IFAC World Congress (IFAC'08)*, Seoul, South Korea, 2008.

[KML08b]   J. Kroneis, P. Müller, and S. Liu. Reduced Order Modeling and Direct Parameter Identification for a Complex Parallel Robot. In *32nd ASME Annual Mechanisms and Robotics Conference (IDETC/MECH'08)*, New York, USA, 2008.

[KML09]   J. Kroneis, P. Müller, and S. Liu. Simplified modelling and parameter identification of parallel robots with redundancies. *Proc. of the Institution of Mechanical Engineers, Part I: J. of Systems and Control Engineering*, 223(1):95–116, 2009.

[Lam03]   Paul Lambrechts. Trajectory planning and feedforward design for electromechanical motion systems. Technical Report DCT 2003-18, Control Systems Technology Group, Faculty of Mechanical Engineering, Eindhoven University of Technology, 2003. Version 2.

[LBS05]   P. Lambrechts, M. Boerlage, and M. Steinbuch. Trajectory planning and feedforward design for electromechanical motion systems. *Control Engineering Practice*, 13(2):145–157, 2005.

[LGZ97]   J. P. Lallemand, A. Goudali, and S. Zeghloul. The 6-dof 2-delta parallel robot. *Robotica*, 15(4):407–416, 1997.

[LHH04]   Qinchuan Li, Zhen Huang, and J.M. Herve. Type synthesis of 3r2t 5-dof parallel mechanisms using the lie group of displacements. *IEEE Trans. on Robotics and Automation*, 20(2):173–180, 2004.

[Lin90]   S.-K. Lin. Dynamics of the manipulator with closed chains. 6(4):496–501, 1990.

[LLL93]   G. Lebret, K. Liu, and F. L. Lewis. Dynamic analysis and control of a stewart platform manipulator. *J. of Robotic Systems*, 10(5):629–655, 1993.

[Lun02]   J. Lunze. *Regelungstechnik 2 - Mehrgrößensysteme, digitale Regelung*. Springer Verlag, Berlin, Heidelberg, New York, 2., neu bearbeitete Auflage, 1. korr. Nachdruck edition, 2002.

[LW03]   Xin-Jun Liu and Jinsong Wang. Some new parallel mechanisms containing the planar four-bar parallelogram. *The Int. J. of Robotics Research,*, 22(9):717–732, Sept. 2003.

[Mö7]   P. Müller. Modellierung und Identifikation von Starrkörper- und Reibungsflüssen am Demonstrator Spidermill, Studienarbeit, Institute of Control Systems, University of Kaiserslautern, 2007.

[Mö8]   P. Müller. Entwurf einer Folgeregelung entlang glatter, zeitoptimaler Trajektorien für die ebene Parallelkinematik SpiderMill, Diplomarbeit, Institute of Control Systems, University of Kaiserslautern, 2008.

[Mar03]   Horacio J. Marquez. *Nonlinear control systems*. John Wiley & Sons, Inc., 2003.

[Mas09]   M. Mas. Full dynamic modelling of the parallel manipulator Spider-Mill using Newton-Euler approach, Diplomarbeit, Institute of Control Systems, University of Kaiserslautern, 2009.

[Mat07]   The MathWorks, Inc. *SimMechanics 2 User's Guide*, 2007.

[Mer00]   Jean-Pierre Merlet. *Parallel Robots*. SOLID MECHANICS AND ITS APPLICATIONS - Volume 74. Kluwer Academic Publishers, Dordrecht / Boston / London, 2000.

[MG92]   D. McFarlane and K. Glover. A loop-shaping design procedure using H_∞ synthesis. *Automatic Control, IEEE Trans. on*, 37(6):759–769, Jun. 1992.

[MLC07]   T. Mbarek, G. Lonij, and B. Corves. Singularity analysis of a fully parallel manipulator with five-Degrees-of-Freedom based on Grassmann line geometry. In *12th IFToMM WC*, pages 1–7, Besançon (France), Jun. 2007.

[Moo04]   Francis C. Moon. *Applied Dynamics - With Applications to Multibody and Mechatronic Systems*. WILEY-VCH, 2004.

[Nai03]   Desineni Subbaram Naidu. *Optimal control systems*. CRC Press, Boca Raton; London; New York; Washington, D.C., 2003.

[Neg03]   R. Negenborn. Robot Localization and Kalman Filters - On finding your position in a noisy world. Master's thesis, Utrecht University, Sep. 2003.

[NH01]   Dan Negrut and Brett Harris. *ADAMS Theory in a Nutshell for Class ME543*. Department of Mechanical Engineering, The University of Michigan, Ann Arbor, Mar. 2001.

[OEC96]   M. Otter, H. Elmqvist, and F. E. Cellier. Modeling of multibody systems with the object-oriented modeling language dymola. *Nonlinear Dynamics 9 (1996)*, pages 91–112, 1996.

[PBBH06]   I. Pietsch, C. Bier, O. Becker, and J. Hesselbach. How to assign time-optimal trajectories to parallel robots - an adaptive jerk-limited approach. *ABCM Symposium Series in Mechatronics, Rio de Janeiro, RJ, Brazil*, 2:174–181, 2006.

[PBKH03]   I.T. Pietsch, O. Becker, M. Krefft, and J. Hesselbach. Time-optimal trajectory planning for adaptive control of plane parallel robots. In *Control and Automation, 2003. ICCA '03. Proc.. 4th Int. Conf. on*, pages 639–643, Jun. 2003.

[PDF91]   F. Pierrot, P. Dauchez, and A. Fournier. Fast parallel robots. *J. of Robotic Systems*, 8(6):829–840, 1991.

[PH05]   F. Pernkopf and M. Husty. Reachable Workspace and Manufacturing Errors of Stewart-Gough Manipulators. *Proc. of MUSME 2005, the Int. Sym. on Multibody Systems and Mechatronics*, pages 203–304, 2005.

[Pie03]   Ingo T. Pietsch. *Adaptive Steuerung und Regelung ebener Parallelroboter*. PhD thesis, TU Braunschweig, 2003.

[PJ86]   F. Pfeiffer and R. Johanni. A concept for manipulator trajectory planning. In *Robotics and Automation. Proc.. 1986 IEEE Int. Conf. on*, volume 3, pages 1399–1405, Apr. 1986.

[PJ87]   F. Pfeiffer and R. Johanni. A concept for manipulator trajectory planning. *IEEE J. of Robotic and Automation*, 3(2):115–123, 1987.

[PK00]   D. Pisla and H. Kerle. Development of dynamic models for parallel robots with equivalent lumped masses. *Int. Conf. on Methods and Models in Automation and Robotics, Międzydroje, Poland*, pages 637–642, 2000.

[PKB⁺05] I.T. Pietsch, M. Krefft, O.T. Becker, C.C. Bier, and J. Hesselbach. How to reach the dynamic limits of parallel robots? An autonomous control approach. *IEEE Trans. on Automation Science and Engineering*, 2(4):369–380, 2005.

[Pro06] Ch. Prothmann. Analyse der Kinematik und Dynamik der parallelkinematischen Struktur Spidermill, Studienarbeit, Institute of Control Systems, University of Kaiserslautern, 2006.

[RAG⁺06] J. M. Rico, L. D. Aquilera, J. Gallardo, R. Rodriquez, H. Orozco, and J. M. Barrera. A more general mobility criterion for parallel platforms. *J. of Mechanical Design*, Volume 128, Issue 1:207–219, 2006.

[Ram00] Rajiv Rampalli. *ADAMS/Solver Theory Seminar*. Mechanical Dynamics, Nov. 2000.

[RH78] M. H. Raibert and B. K. P. Horn. Manipulator control using the configuration space method. *The Industrial Robot, June 1978*, pages 69–73, 1978.

[Rit05] Th. Rittenschober. Entwicklung einer Fahrdynamikregelung am virtuellen Fahrzeug - Einsatz aktiver Vorder- und Hinterradlenkung und Bremse. In *Internationales Forum Mechatronik 2005*, Augsburg, Germany, Jun. 2005.

[Rus88] R. Russell. Kinematic optimization of lower-pair clutch mechanisms. Technical report, M. Eng. Project, Dept. Mech. Eng., McGill University, Montréal, Québec, Canada, 1988.

[Sch03] M. Schlotter. *Multibody System Simulation with SimMechanics*, May 2003.

[SD85] Z. Shiller and S. Dubowsky. On the optimal control of robotic manipulators with actuator and end-effector constraints. In *Robotics and Automation. Proc.. 1985 IEEE Int. Conf. on*, volume 2, pages 614–620, Mar. 1985.

[Sha05] Ahmed A. Shabana. *Dynamics of Multibody Systems*. Cambridge University Press, 2005.

[Shi94] Z. Shiller. On singular time-optimal control along specified paths. *Robotics and Automation, IEEE Trans. on*, 10(4):561–566, Aug. 1994.

[SHV06]   Mark W. Spong, Seth Hutchinson, and M. Vidyasagar. *Robot Modelling and Control*. John Wiley & Sons, Inc., 2006.

[Sim06]   Dan Simon. *Optimal State Estimation - Kalman, H_∞, and Nonlinear Approaches*. John Wiley & Sons, Inc., Hoboken, New Jersey, 2006.

[Sim08]   S. Simon. Entwurf und Implementierung von dezentralisierten und zentrallisierten Regelungskonzepten für die Parallelkinematik SpiderMill, Studienarbeit, Institute of Control Systems, University of Kaiserslautern, 2008.

[SK08]   Bruno Siciliano and Oussama Khatib, editors. *Springer Handbook of Robotics*. Springer-Verlag Berlin Heidelberg, 2008.

[SL88]   J.-J.E. Slotine and Weiping Li. Adaptive manipulator control: A case study. *Automatic Control, IEEE Trans. on*, 33(11):995–1003, Nov. 1988.

[SL91]   J.-J. E. Slotine and W. Li. *Applied Nonlinear Control*. Prentice-Hall, Inc., 1991.

[SL92]   Zvi Shiller and Hsueh-Hen Lu. Computation of path constrained time optimal motions with dynamic singularities. *J. of Dynamic Systems, Measurement, and Control*, 114(1):34–40, 1992.

[SM83]   Kang G. Shin and Neil D. Mckay. An efficient robot arm control under geometric path constraints. In *Decision and Control, 1983. The 22nd IEEE Conf. on*, volume 22, pages 1449–1457, Dec. 1983.

[SM85]   Kang Shin and N. McKay. Minimum-time control of robotic manipulators with geometric path constraints. *Automatic Control, IEEE Trans. on*, 30(6):531–541, Jun. 1985.

[SMB07]   S.-D. Stan, V. Maties, and R. Balan. Workspace analysis of the Biglide mini parallel robot with 2 DOF. In *Proc. IEEE Workshop on Advanced Robotics and Its Social Impacts ARSO 2007*, pages 1–6, 2007.

[SP05]   Sigurd Skogestad and Ian Postlethwaite. *Multivariable Feedback Control-Analysis and Design*. Wiley, 2005.

[SR97]   Lawrence F. Shampine and Mark W. Reichelt. *The MATLAB ODE Suite*. SIAM J. Sci. Comp., 1997.

[SRK99]   L. F. Shampine, M. W. Reichelt, and J. A. Kierzenka. Solving Index-1 DAEs in MATLAB and Simulink. *SIAM Rev.*, 41(3):538–552, 1999.

[SS05]   L. Sciavicco and B. Siciliano. *Modelling and Control of Robot Ma-*

nipulators. Springer, 2nd ed., 6th printing edition, 2005.

[SSdc71] N.K. Sinha, B. Szabados, and C.D. di cenzo. New high-precision digital tachometer. *Electronics Letters*, 7(8):174–176, 22 1971.

[SSVO09] Bruno Siciliano, Lorenzo Sciavicco, Luigi Villani, and Giuseppe Oriolo. *Robotics - Modelling, Planning and Control*. Springer-Verlag London Limited, 2009.

[ST98] Richard Stamper and Lung-Wen Tsai. Dynamic modeling of a parallel manipulator with three translational degrees of freedom. Atlanta, Georgia, Sept. 13-16 1998.

[Sta97] R. E. Stamper. *A Three Degree of Freedom Parallel Manipulator with Only Translational Degrees of Freedom*. PhD thesis, University of Maryland, 1997.

[SW04] Heinz-Bodo Schmiedmayer and Martin Weigel. ADAMS - Ein Simulationspaket für Mehrkörpersysteme. *ZIDLINE*, (10), Jun. 2004.

[SY89] J.-J. E. Slotine and H. S. Yang. Improving the efficiency of time-optimal path-following algorithmus. *IEEE Trans. on Robotics and Automation*, 5(1):118 – 124, 1989.

[TA81] Morikazu Takegaki and Suguru Arimoto. A new feedback method for dynamic control of manipulators. *J. of Dynamic Systems, Measurement, and Control*, 103(2):119–125, 1981.

[Tom91] P. Tomei. Adaptive PD controller for robot manipulators. *Robotics and Automation, IEEE Trans. on*, 7(4):565–570, Aug 1991.

[Tom00] P. Tomei. Robust adaptive friction compensation for tracking control of robot manipulators. *IEEE Trans. on Automatic Control*, 45(11):2164–2169, 2000.

[Tsa99] Lung-Wen Tsai. *Robot Analysis - The Mechanics of Serial and Parallel Manipulators*. John Wiley & Sons, Inc., 1. edition, 1999.

[Wal66] K. J. Waldron. The constraint analysis of mechanisms. *The J. of Mechanisms*, 1:101–114, 1966.

[WB06] Greg Welch, , and Gary Bishop. An Introduction to the Kalman Filter. Technical Report TR 95-041, Department of Computer Science, University of North Carolina at Chapel Hill, Chapel Hill, 2006.

[WG07] Yangnian Wu and Clément M. Gosselin. On the dynamic balancing of multi-dof parallel mechanisms with multiple legs. *J. of mechanical design*, 129(2):234–238, 2007.

[WH88]   K. J. Waldron and K. H. Hunt. Series-parallel dualities in actively coordinated mechanisms. In *Proc. of the 4th Int. Symposium on Robotic Research*, pages 175–181, Cambridge, MA, 1988. MIT Press.

[WK03]   Giles D. Wood and Dallas C. Kennedy. *Simulating Mechanical Systems in Simulink with SimMechanics*. The Math Works, 3 Apple Hill Drive, Natick, MA, USA, 2003.

[WSO02]   G. J. Wiens, S. A. Shamblin, and Y. H. Oh. Characterization of pkm dynamics in terms of system identification. *Proc. of the Institution of Mechanical Engineers, Part K: J. of Multi-body Dynamics*, 216(1):59–72, 2002.

[XKW92]   Y.-X. Xu, D. Kohli, and T.-C. Weng. Direct differential kinematics of hybrid-chain manipulators including singularity and stability analyses. In *22nd Biennial Mechanisms Conf.*, volume DE-45, pages 65–73, Scottsdale, Sept., 13-16 1992.

[YK00]   Koji Yoshida and Wisama Khalil. Verification of the positive definiteness of the inertial matrix of manipulators using base inertial parameters. *The Int. J. of Robotics Research*, 19(5):498–510, 2000.

[Ż03]   S. H. Żak. *Systems and Control*. Oxford University Press, Inc., 2003.

[ZBG02]   Dimiter Zlatanov, Ilian A. Bonev, and Clément M. Gosselin. Constraint singularities of parallel mechanisms. *Proc. of the IEEE Int. Conf. on Robotics and Automation, Washington, DC*, pages 496–502, 2002.

[ZDG96]   Kemin Zhou, John C. Doyle, and Keith Glover. *Robust and optimal control*. Prentice-Hall, Inc., Upper Saddle River, NJ, USA, 1996.

[ZH00]   T. S. Zhao and Z. Huang. Theory and application of selecting actuating components of spatial parallel mechanisms. *Chinese J. of Mechanical Engineering*, 36(10):81–85, 2000.

[ZH07]   Si-Jun Zhu and Zhen Huang. Eighteen fully symmetrical 5-dof 3r2t parallel manipulators with better actuating modes. *The Int. J. of Advanced Manufacturing Technology*, 34(3-4):406–412, 2007.

[Zha10]   J. Zhao. Vergleich linearer und nichtlinearer Beobachtungskonzepte zur Schätzung der Geschwindigkeit der aktiven Gelenke des Demonstrators SpiderMill. Master's thesis, Institute of Control Systems, University of Kaiserslautern, 2010.

[ZP02]   Yongliang Zhu and P.R. Pagilla. Static and dynamic friction compensation in trajectory tracking control of robots. In P.R. Pagilla, editor, *Proc. IEEE Int. Conf. on Robotics and Automation ICRA '02, Washington, DC, USA*, volume 3, pages 2644–2649, 2002.

[ZS93]   Chang-De Zhang and Shin-Min Song. An efficient method for inverse dynamics of manipulators based on the virtual work principle. *J. of Robotic Systems*, 10(5):605–627, 1993.